Philosophical Biology in Aristotle's Parts of Animals

STUDIES IN HISTORY AND PHILOSOPHY OF SCIENCE

VOLUME 26

For further volumes:
http://www.springer.com/series/5671

Jason A. Tipton

Philosophical Biology in Aristotle's Parts of Animals

Springer

Jason A. Tipton
St. John's College
Annapolis, MD, USA

ISSN 0929-6425
ISBN 978-3-319-37652-3 ISBN 978-3-319-01421-0 (eBook)
DOI 10.1007/978-3-319-01421-0
Springer Cham Heidelberg New York Dordrecht London

Printed on acid-free paper

Springer is part of Springer Science+Business Media (www.springer.com)

Preface

If my life were two dimensional, this book would have seen the light of day years ago. I am so grateful to my wife Liz and our children, Zach and Nora, for adding a third dimension. There have been many times when Liz watched from the shore as I spent countless hours trying to catch some nondescript fish as part of the work that went into this book. She has been extremely good-natured as I have tried to coax our children into such mad pursuits.

Acknowledgments

The seeds of this book were sown in the office of my mentor, Ronna Burger. Given my interest in biology, she could not understand why I even contemplated working on Plato as a dissertation project, as I had done for my master's thesis. I naturally belonged among Aristotle's more biological works. Ronna not only gave me that initial push but challenged my thinking at every step. Her gentle influence is clear to me in most of what is here.

I am also particularly grateful to my compatriots in philosophical study at Tulane University—Steven Berg, Michael Golluber, and Matt Oberrieder. Henry Bart, my advisor in Tulane University's Ecology and Evolutionary Biology Department, helped further my love of the biological world, especially the world of fishes. Conversations with colleagues and students at St. John's College, Annapolis, helped sharpen my thinking about Aristotle.

I was very fortunate to receive a Fulbright Fellowship to Greece during the 2000–2001 academic year as this project was coming into focus for me. In addition, Sherry Fox and the Wiener Lab at the American School of Classical Studies in Athens (ASCSA) helped in many ways as I attempted to follow Aristotle. The ASCSA, a place more frequently populated with archaeologists and classicists, provided a warm welcome to me and is the source of many fond memories. The people of Lesvos, Greece, offered help and advice at many junctures. Among the Lesvians, Hjalmar Dahm and Georgios Sitaras offered invaluable logistical assistance and field help.

Contents

Chapter 1
Aristotle's Philosophy and Biology: The Biological Phenomena

Abstract What are we to do with the wealth of detailed information in the biological works of Aristotle? How easy is it to clearly distinguish between what some might describe as "merely" biological and the more philosophical, speculative discussions? Can the activity in which Aristotle is engaged be described as a philosophical biology? What would such an inquiry entail? This book aims to examine these questions through a detailed analysis of Aristotle's *Parts of Animals* in conjunction with revisiting the detailed natural history observations made by Aristotle that inform, and in many ways penetrate, the philosophical argument.

In the scholarly works within the history and philosophy of biology, one does not often find data, illustrations or photographs of organisms. One can, say, certainly read the zoological works of Aristotle without reference to the parts or animate objects discussed in those writings. In the same way, one can read the *Poetics* without having read a particular tragedy under consideration. But I tend to want to have the biological phenomena in front of me when reading Aristotle's *Parts of Animals (HA)*. For me, the phenomena add an element to the argument; perhaps I am convinced of Aristotle's appeal to the Heraclitean idea that "there are gods here too" when exhorting his readers to look into what one might consider low and insignificant (645a5–a35). Is there something divine about cephalopod mouths or fish tongues? While undoubtedly his exhortation is at least mildly hyperbolic, I am convinced by him to attempt to examine lowly organic forms in the hope that they help me better understand his thinking. D'Arcy Thompson, makes a more general claim when he suggests that Aristotle "recognized the great problems of biology that are still ours to-day, problems of heredity, of sex, of nutrition and growth, of adaptation, of the struggle for existence, of the orderly sequence of Nature's plan. Above all he was a student of Life itself. If he was a learned anatomist, a great student of the dead, still more was he a lover of the living. Evermore his world is in movement" (Thompson 1913, p. 15).

J.A. Tipton, *Philosophical Biology in Aristotle's Parts of Animals*, Studies in History and Philosophy of Science 26, DOI 10.1007/978-3-319-01421-0_1,

1.1 The Biological Phenomena

In the introduction to his excellent commentary on the *PA*, James Lennox (2001a) states that his is a primarily philosophical "rather than 'scientific'" commentary (p. xiii).[1] I aim to show that an interpretation of the *PA* can be both philosophical and scientific. The scientific aspects of the *PA* are one way into the philosophical aspects of the inquiry. Ultimately, I want to explore the possibility that what looks like parallel paths—one philosophical, one scientific—can be interwoven in a way that Aristotle seems to have intended.

The analysis of the philosophical argument of the *PA* will be enhanced by appealing to the actual organisms and traits that were of interest to Aristotle as a philosophical biologist. If one is suspicious about the veracity of certain of Aristotle's observations, one might be suspicious of the philosophical argument. This might be what David Balme worries about when he claims:

> Much of this criticism [of these biological writings] arose in the nineteenth and early twentieth centuries from armchair naturalists who disbelieved Aristotle's reports and thought them too silly for a great philosopher... I confess that I was still blaming Aristotle for swallowing the story about buffaloes projecting their dung at enemies, until in 1983 I saw a picture on television of hippopotamuses doing just that. (1987a, pp. 16–17)

In a similar vein, Lennox's (1983) examination of echinoderm reproductive parts is so illuminating because of its careful analysis of the empirical side of Aristotle. A number of Aristotle's observations about particular aspects of the anatomy, physiology, and adaptation of organisms have only been recently rediscovered, e.g. the reproductive behavior of the pipe-fishes (Sygnathidae), the hermaphroditism of members of the Serranidae, and the nest building behavior of wrasses (Labridae) (Thompson 1913).[2] Buddington and Diamond (1986) revisited Aristotle's ruminations on the function of the pyloric caeca in fish. Brock (2004) pointed out that Aristotle's discussion of sperm competition in birds is largely accurate, despite the scathing remarks of the translators and commentators. In examining Aristotle's understanding of marine invertebrates, Voultsiadou and Vafidis (2007) remark that "the great philosopher had a remarkable, well-balanced scientific knowledge of the diversity of the various invertebrate groups, very similar to that acquired by modern marine biologists in the same area of study" (p. 103). These are what we might categorize as more scientific commentaries, to use Lennox's formulation. And I have certainly completed work in this vein (Tipton 2006, 2008). This book will attempt to hold together what might look like, at first glance, the two aspects of the *PA*: our understanding is enhanced by a detailed analysis of the organisms and traits that consumed Aristotle; it is in this that the outlines of a philosophical biology begin to emerge.

---

[1]Lennox (2001) describes much of the scholarly work that has been done on Aristotle's *PA*.

[2]These are all different behavioral or physiological aspects of different fishes.

Fig. 1.1 D'Arcy Thompson (1913, p. 13) says that "I take it then as probable, or even proven, that an important part of Aristotle's work in natural history was done upon the Asiatic coast, and in and near to Mitylene. He will be a lucky naturalist who shall go some day and spend a quiet summer by that calm lagoon, find there all the natural wealth *hosson Lesbos . . . entos eeryei*, and have around his feet the creatures that Aristotle loved and knew." Lesvos is a relatively large island (163,000 ha) in the Northern Aegean Sea. The Bay of Kalloni is a large (14,500 ha) body of water that roughly divides the island into two large lobes. A large marsh area (*below*) characterizes the head of the Bay near Pyrrah

The bulk of Aristotle's work in natural history was done in the North Aegean, centered around Mitylini on the north Aegean island of Lesvos (Thompson 1913; Lee 1985). In fact, Lesvos and Pyrrah (on the Bay of Kalloni) are among the very few place names mentioned by Aristotle (e.g. *HA* 621b12, 621b22; Fig. 1.1). It is generally acknowledged that Aristotle traveled to Lesvos with Theophrastus, who was from Eressos, shortly after Plato's death.

I have spent 2 years in Lesvos, the location of much of Aristotle's work in natural history, familiarizing myself with the creatures that Aristotle had at his finger-tips. I have done this while trying to keep an eye on the philosophical discussion of the *PA*.

Fig. 1.2 Ascidians or sea-squirts grow in potato-like clumps. These were collected in the Gulf of Gera, Lesvos. Ascidians are characterized by two orifices, inhalant and exhalant siphons (*arrows, upper right*). The difficulty in identifying one of the orifices as mouth and other as functioning for the sake of discharge is perplexing to Aristotle in examining these plant-like animals. However, the presence of these orifices allows Aristotle to suggest that ascidians are more animal-like in their nature than are sponges (*PA* 681a10). On the other hand, the ascidians appear to have no residue, despite having an orifice that looks as if it is for waste disposal, which makes them plant-like (*PA* 681a31). The second orifice is for the sake of discharging water (*HA* 528a14). The ascidians are testaceans, or hard-shelled organisms. But unlike things like oysters, ascidians are enclosed in leathery husk (*HA* 528a2). These animals, especially when handling them, do not immediately go together with the oysters. But, as Aristotle says, "compared with one another testacea exhibit many differences, in respect both of their shells and their flesh within" (*HA* 528a5). The leathery husk has to be understood as a shell which "completely envelops their flesh" (*HA* 528a21). The texture of the shell is between that of skin and shell (531a11)

Again, what does handling such things or even inspecting illustrations of such phenomena add to our understanding of the argument? At the very least, a photograph or illustration can bring to life what it is Aristotle is talking about. More significantly, seeing the phenomenon might allow one to better understand why Aristotle makes a certain argument or interpretation.

For example, do readers of Aristotle have in mind organisms like sea-squirts (ascidians) or sea cucumbers (holuthurians) when trying to understand Aristotle's argument regarding plant-like animals (Figs. 1.2 and 1.3)? It has been my own

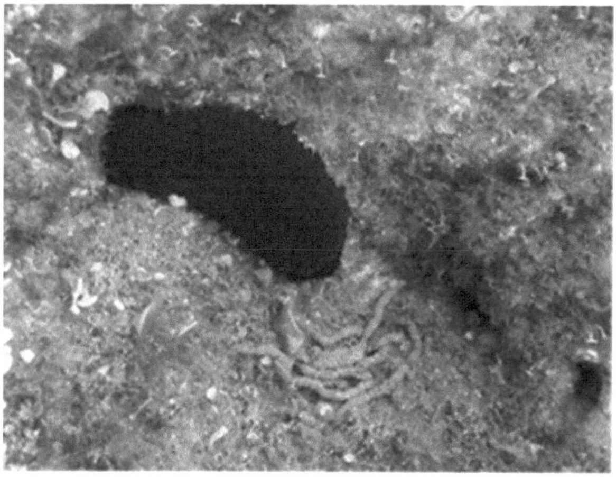

Fig. 1.3 A sea cucumber or holothurian on the sea floor, in approximately 3 m of water. Note the fecal pellets trailing behind, marking both the slow movement and the processing of nutriment. The observation of such fecal pellets made Aristotle more certain that organisms like these, while still identified as plant-like, were more confidently identified with the animals. While the sea-cucumber is not attached like a sponge, it is generally stationary (*HA* 487b15)

Fig. 1.4 Aristotle's *kobios* (*Gobius cobitis*)

experience that I understand the terms of the philosophical argument in a richer way if I have the phenomena in front of me. The discussion of plant-like animals is important in Aristotle because of the question about the continuum between plant and animal life. Where does Aristotle draw the line? Plant-like animals bring this question into focus and demonstrate the indeterminacy of any potential solution to the division.

To give another example close to me, I spent 1 year trying to better understand Aristotle's observations of two small, rather insignificant fishes, his *kobios* (*Gobius cobitis*) (Fig. 1.4) and *phucis* (*Parablennius sanguinolentus*) (Fig. 1.5). I discovered that Aristotle's descriptions of their habitat, feeding behavior, spawning and other attributes were largely accurate and understood better, after studying them myself, the sensual experiences that might have contributed to Aristotle's discussion of

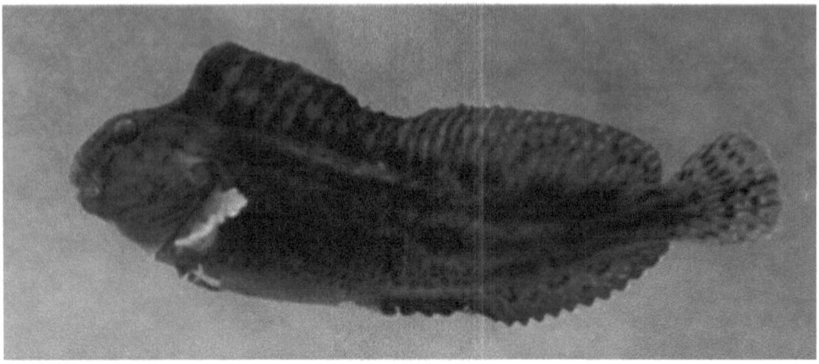

Fig. 1.5 Aristotle's *phucis (Parablennius sanguinolentus)*

them. These small, cryptobenthic fishes share much in common. Their coloration is a very similar salt and pepper mottling. *Gobius cobitis* and *P. sanguinolentus* have similar body shapes, with relatively robust head regions that taper towards a rounded tail. They have upturned eyes and terminal mouths. The gobies have modified pelvic fins that are united into a sucking disc to adhere to substrate. *Parablennius sanguinolentus* has very pronounced pelvic fin rays that seem to allow it to maintain a position on the substrate. When Aristotle discusses the *echeneis*, which is usually thought to be a blenny, a goby or a remora, he says that its fins resemble feet (*HA* 505b20), which could also be said of *P. sanguinolentus*. *Gobius cobitis* and *P. sanguinolentus* have very different looking pelvic fins, but they are similar in their work or activity; the pelvic fins of both kinds are such that they suit a lifestyle oriented towards the bottom, towards the substrate. Aristotle would suggest that they are functionally analogous, which would help us understand why he talks of the *kobios* and *phucis* in the same contexts. The pelvic fins of both are well suited for the very turbulent conditions that occur when the winds are persistently strong across the Bay of Kalloni. As similarly functioning beings embedded in the same environment, the fishes are morphologically similar and have many overlapping characters and ways. Pelvic fins which might be said to differ by means of "the more and the less" serve the same function. We see in this small example the way in which the work of an organic whole can be accomplished in various ways.

The fact that *kobios* and *phucis* are hard to distinguish as distinct kinds in Aristotle's writing, I would suggest, has everything to do with their common habits of life, with the powers that each kind has in the struggle for existence.

Let me mention another example highlighting the level of detail to be found in Aristotle. He makes really remarkable observations regarding what contemporary ichthyology would describe as a feeding heteroaggregation, the co-foraging of individuals of different species; Aristotle described the feeding of the red mullet (*Mullus spp.*) and sea bream (*Diplodus spp.*) in terms that would be recognizable to biologists today (see Tipton 2008 for a detailed discussion of the particulars). This kind of co-foraging of the red mullet and sea bream is remarkable. What seems to me equally remarkable is that Aristotle recorded this aspect of the life history of these fish. How did he, or his source, make such observations? Did they observe the sandy bottom from shore?

While I have observed different fish from shore, noting their movements and behaviors in reference to Aristotle's own work (Tipton 2006), it is difficult for me to imagine someone observing this cooperative foraging from shore. Did they see these fish interact in cisterns set up to hold fish (Fig. 1.6)? Given its status as a delicacy, one could easily imagine considerable resources being devoted to the maintenance of a fresh supply of red mullet. Pliny described the extraordinary efforts made in the construction and maintenance of oyster and fish ponds (Rackham 1940, p. 277). Perhaps if they were in such an enclosure or pond of some kind, Aristotle or his source could more easily observe their behaviors. Could it be that this remarkable behavior was observed and noted while diving? Anyone who has swum without a mask could report that it would be very difficult to see with enough clarity to make the detailed observations of *Mullus* species foraging in even 3 or 4 m of water. Perhaps Greek divers had a capacity or a technique that allowed such a feat. In his thoughts on ancient diving, Frost (1968) makes the following observation:

> Both Minoan and Mycenaean graphic arts reveal a lively appreciation for undersea life: octopuses, shellfish, sponges, and their rocky environment are depicted with charm and a high degree of accuracy as well. Sponges, it might be added, are almost never identified as such by modern commentaries on vase paintings, but anyone who has dived into the Aegean recognizes them immediately for what they are (p. 181).

Aristotle does describe some of the equipment that might have been available to divers during his time. To a modern reader, Aristotle seems to give a description of a device which resembles a snorkel: "Some divers, when they go down into the sea, provide themselves with a breathing machine, by means of which they can inhale the air from above the surface while they remain for a long time in the water" (*PA* 659a8–13). He also described the way in which a cauldron can be lowered into the water, upside down, keeping some air inside which could be utilized by a diver, a sort of diving bell (*Problems* 960b32).

Both of these devices could aid a diver in antiquity but would do nothing to correct that individual's extremely blurry vision under the water. With this problem in mind, it is interesting to note that lenses made of rock crystal have been discovered at archaeological sites across the Mediterranean (James and Thorpe 1994, pp. 157–163). They are thought to have been useful as an aid to the eyes in doing detailed engraving and sculpting. While such lenses could be used to magnify or concentrate the rays of the sun to burn—as Strepsiades described in Aristophanes' *Clouds*—I wonder if such things could have utilized as a crude mask for diving. The detailed description of the feeding habits of the red mullet and the striped red mullet might suggest the use of something like a mask. If what Aristotle described was a kind of snorkel-like instrument, one might speculate on the existence of an accompanying mask.

In the description of the interaction between the bream and red mullet, Aristotle observed that the former can take advantage of the work of the latter. That work includes digging or stirring up of the mud, which is made possible by means of

Fig. 1.6 An ancient holding tank for fish erected near the south harbor in Mytilini, Greece and dates from the first century CE. The holes seen one third of the way up the wall are thought to be hiding places and habitat niches for fish and invertebrates (Hector Williams, per. comm. July 2005)

the chin barbels. Aristotle not only provided many aspects of the life history of the red mullet and sea bream, he also described this peculiar interaction, described as heterospecific aggregation by contemporary researchers (for example, De Pirro et al. 1999) studying some of the same species and problems today.

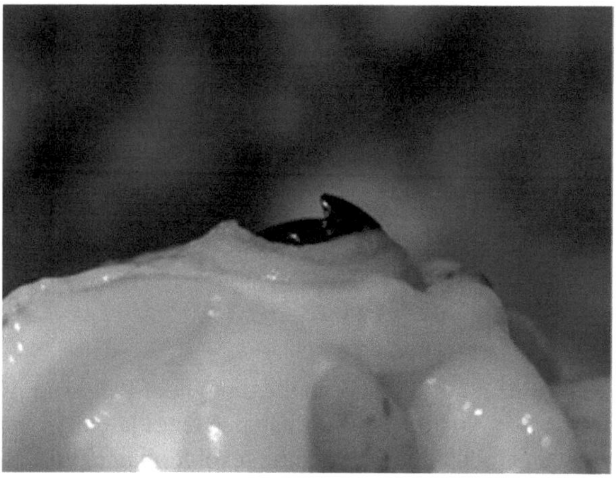

Fig. 1.7 The mouth of a sepia or cuttlefish looks like a bird beak

The extremely peculiar particulars of the red mullet/bream feeding interaction are one example that points to a larger theme. Aristotle's discussion of the common activity of feeding might be another more general example that helps illustrate the way in which the biological details might illuminate the larger argument. All animals have a mouth and an anus. A mouth then, we could consider, is associated with the most common actions (*ta praxeis*) of animals. Aristotle is interested in mouths—mouths of fishes, of birds, of pigs, of bloodless things like insects and marine invertebrates. But what is included in a mouth? Lips, tongue, teeth? In looking at birds, Aristotle suggests that lips and teeth are fused to form a beak. In looking at cephalopods, he sees two teeth (*PA* 678b7) and something analogous to a tongue. But what Aristotle calls teeth, one might see in a cephalopod mouth what looks like a beak (Fig. 1.7); one better understands why Aristotle calls them teeth when one sees how they are surrounded by a mouth or lip-like structure. But cephalopod teeth are like bird beaks, according to Aristotle, in being inefficient for grinding food, thus necessitating a crop. There are similarities and differences between cephalopods and birds in this most necessary of parts. One sees in this the way in which we must appeal to the necessary cause, as well as the final one, in recognizing that the mouths of both cephalopods and birds fall short in some way. I have chosen these examples to highlight what I hope to accomplish because they might be viewed as low or trivial. But, these two cases—the case of the *kobios* and the discussion of cephalopod mouth parts—inform the argument regarding issues such as division of kinds and necessity in a causal account.

This book project aims to open and develop a new avenue of research into the recently renewed interest in Aristotelian natural philosophy. In orienting this project with an eye to both the empirical and philosophical, it is hoped that this book will offer a way to begin to understand better Aristotle's philosophical biology at the same time as trying to understand his biological philosophy. In this interpretive work I do not offer here an exhaustive catalogue of the parts or organisms spoken of in the *PA*; such an undertaking would take many years not to mention serious consideration of the *History of Animals* (*HA*). I am not an expert in invertebrate zoology, ichthyology, or ornithology, nor am I a very good photographer; but instead I hope to combine a thoughtful reading of Aristotle's *PA* with a curious eye directed to Aristotle's animated world.

1.2 Aristotle's "Pre-Socraticism"

Though there are significant differences in the thought of Plato and Aristotle, I start with the assumption that Aristotle was an independent Platonist,[3] that he was part of the tradition that emerged after the Socratic revolution in philosophy. But as will become immediately clear, this assumption makes manifest certain serious problems and may appear naïve. When Phaedrus accuses Socrates of being somehow out of place outside of the city, Socrates responds by saying, "For I am a lover of learning. Now then, the country places and the trees are not willing to teach me anything, but the human beings in town are" (*Phaedrus* 230d). It looks as if Socrates is content to restrict his own philosophical thinking to the human things, to the things of the city, whereas Aristotle appears to have learned a great deal indeed from the trees and other non-human organisms outside of the city walls.

The picture Socrates paints of himself in the *Phaedrus* was the product of a revolution in his thinking, which he describes in his intellectual autobiography in Plato's *Phaedo*. Socrates refers there to his early investigations as a kind of inquiry into nature, a *historia peri phuseos*. From a well-known formulation by Cicero, it is generally acknowledged that Socrates brought philosophy down from the heavens and made it dwell in the city. Socrates describes that move as a turn to *logoi*. It is a turn away from something else, or someone else: it is as if he turns away from the Pre-Socratic Socrates, who seems to be the one represented comically by Aristophanes.

The Pre-Socratic philosophers, whom Aristotle calls the first philosophers, are those who "discourse on nature" or *phusis*. *Phusis* refers primarily to the sphere of things that grow, as opposed to things made, like a chair. The recognition of *phusis* as a distinctive sphere did not always exist. Before *phusis* was discovered, all things were explained as having a way or custom.[4] So we could say that the dog

---

[3]David Sedley (2010) emphasizes "Aristotle's Platonic background and training" (p. 5). "Independent Platonist" is Sedley's forumulation. It is likewise fruitful to locate Aristotle within the larger Socratic project.

[4]How was the notion of *phusis* discovered? The term "*phusis*" occurs exactly once within Homer. In book 10 of the *Odyssey*, Odysseus recounts how Circe turns his companions into swine. Specifically, she causes them to look like pigs while maintaining some aspects of their human minds. While venturing to Circe's house, Odysseus is approached by Hermes, who gives him a drug called moly to counter the effects of any potions Circe might attempt to use on him. Hermes

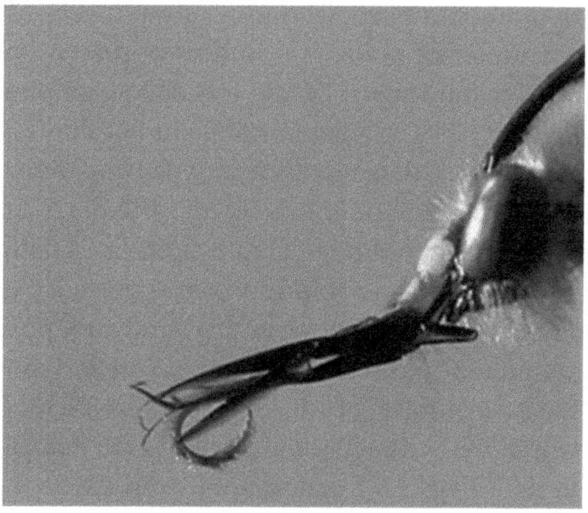

Fig. 1.8 Close up detail of a bee's mouth

has a certain way, just as a certain tribe of people might have their way. With the discovery of *phusis*, the way of something becomes divided into two, into *phusis* and *nomos*, law or custom.

The study of nature occupies a central place in Aristotle's thought. Of the Aristotelian corpus, a significant portion is devoted to what are collectively referred to as the "biological writings," such as the *History of Animals*, the *Parts of Animals*, and the *Progression of Animals*. In these works, at least, Aristotle seems to be much closer to the Aristophanic image of Socrates than to the Platonic image of the mature Socrates, discussing courage or moderation or other such things with various interlocutors. Is Aristotle, then, in his study of nature, simply going back behind the Socratic turn and taking up again the pre-Socratic project? Or is it possible that his exploration of the natural world is in some way a continuation of the Socratic turn?

There are not many areas of the organic world that Aristotle did not examine and think about, for "in all natural-things (*phusikois*), there is something wonderful (*thaumaston*)" (*PA* 645a17). Aristotle's enthusiasm for studying the organic world led him to some peculiar subjects. In commenting on his description of the parts of the mouths of insects, Ogle praises him as follows: "It is evident that Aristotle must have examined with much care the oral arrangements of insects; and when we consider how difficult it is without magnifying glasses to make them out, we cannot but be struck with admiration at his considerable success in the matter" (note 8, p. 220; Fig. 1.8). While appreciating Ogle's praise, we can also note that

explains the nature (*phusis*) of the moly to Odysseus, telling him that it is black at the root, but with a milky flower. This description is to give the nature of the thing in terms of its look. In addition to this nature, the moly has the power of medicine for human beings, in this case the power (*dunamis*) to counter the potions of Circe. Odysseus will be protected from Circe by the power of the moly. This description highlights the power that certain organic things have; in this context, the power is interpreted from the perspective of the human use. The power of the moly root is a power or *dunamis* insofar as it has a function. Naddaf (2005) examines this passage as well as the pre-Socratics, most notably Anaximander, to better understand their interest in *phusis* and, more surprisingly given certain caricatures of the those earlier thinkers, their interest in political life.

studying the inner workings of the insect world can also be portrayed as comical: Aristophanes portrays a student of the *phrontisterion* recounting to Strepsiades the Socratic musings on fleas' feet and gnats' rumps (*Clouds* 140–167). Such biological inquiries can be seen as comical; but they are, in Aristotle's eyes, somehow a way into a world that is both wonderful and divine (*PA* 645a15–24).

The perplexities raised by the attempt to understand plant and animal life open up fundamental ontological questions, as Aristotle indicates throughout his biological writings. At the same time, he repeatedly turns to biological examples to clarify questions encountered in the study of physics or metaphysics, for "plants and animals are substances or beings (*ousiai*) most of all" (*Meta*. Z.8). If organic things are beings most of all, it seems that Aristotle, in his investigation of the organic world, looks directly at the beings, the beings that Socrates looked at only indirectly in the *logoi.* As Lennox points out (2001, p. 121), *PA* I " makes use of this term [*ousia*] in a variety of ways, and is one of the more interesting texts in which to explore its meaning." As we shall see in this analysis of the *PA*, the study of the nature of the organic world was Aristotle's way into such ontological problems as the relationship between matter and form, or form and function, or the heterogeneity of the many different kinds of being.[5]

But we must wonder about the connection between the study of nature and the study of being. Aristotle suggests that his Pre-Socratic predecessors did not have the notion of substantial being (*ousia*)[6] and even though Democritus touched upon the notion, it was not thought to be necessary for the study of nature (642a27). It was Socrates, Aristotle seems to suggest, who is the discoverer of the study of being: his turn to the human and political things was for the sake of a study of being (642a29). The description of the Socratic turn Aristotle provides in the *Metaphysics* (987b)—as a project restricted to the human things—begins to look like a caricature. Plato's Socrates did not present his discovery that way. Rather, Socrates describes his decisive turn as a new way of proceeding: in place of his predecessors' attempt to apprehend the beings directly, he 'took refuge in *logoi'* and followed the indirect path of examining speeches (99d–e). Plato's Socrates, unlike Aristotle's, does not restrict his territory to "the human things".[7] Aristotle inherited the study of ontology from the Socratic tradition and his study of nature, although, at first blush, appears to be in tension with Socratic philosophy, is in the service of ontological explorations. Aristotle's project in his biological writings can be understood, then, as an attempt to mine an unexplored vein in the Socratic tradition, investigating the non-human animal world with the fundamental questions about the nature of being in mind.

With regard to the question of the connection between metaphysics and biology, Montgomery Furth (1987) even goes so far as to suggest that much of Aristotle's

---

[5]This stance puts me in the camp that sees findings in his biological work as contributing significantly to his metaphysics. Bolton (2010) offers a robust argument against this view, arguing against those who see biology and metaphysics as mutually illuminating.

[6]I will generally try to follow Lennox in translating *ousia* as "substantial being."

[7]I am indebted to Ronna Burger in understanding this difference.

metaphysics of substance is motivated by an attempt to provide a theoretical foundation for the study of biology. In the last 50 years there has been a tremendous renewal of interest in Aristotle's biological writings, interest motivated by at least two different sorts of researchers who had distinct assumptions and motivations. On the one hand there are the biologists, and chief among them, D'Arcy Thompson.[8] Even Darwin was impressed with Aristotle the biologist.[9] On the other hand there are those who, in their study of Aristotle's philosophic work, have been forced to recognize that a significant portion of the corpus had been neglected; these scholars wish to make sense of how Aristotle's biological studies are to be understood in light of the canonical works, e.g. the *Metaphysics*, *Physics*, *De Anima*, *Nichomachean Ethics*.

Let me suggest how this study of the *PA* will contribute to recent scholarship, while attempting to illuminate certain ontological themes through the discussion of zoology. There is a consensus among most scholars that in *PA* I, Aristotle recognizes the need to incorporate explanations based on necessity and teleology, as a methodological point. However, an analysis of the movement of the argument of *PA* I makes clear that another methodological point emerges.[10] In a discussion near the end of *PA* I of the way in which the study of nature is to be conducted,

---

[8]The problems that Aristotle confronts are the same, in many cases, as those addressed by modern biological thought. Consider D'Arcy Thompson's claim that Aristotle "recognized the great problems of biology that are still ours to-day, problems of heredity, of sex, of nutrition and growth, of adaptation, of the struggle for existence, of the orderly sequence of Nature's plan. Above all he was a student of Life itself. If he was a learned anatomist, a great student of the dead, still more was he a lover of the living. Evermore his world is in movement" (1913, p. 15). While Thompson may overstate the common ground between modern biological thought and Aristotle's own thought—especially in using Darwinian formulations like the "struggle for existence"—students of Aristotle can recognize how Thompson might make such a claim. However we might view Thompson's attitude toward Aristotle, it is surely closer to the mark than the eminent anatomist G.G. Simpson's (1961, 36n.) alleged agreement with Bacon: "I tend to agree with Roger Bacon that the study of Aristotle increases ignorance."

[9]Compare Darwin's appraisal of Aristotle: "Linnaeus and Cuvier have been my two gods . . . but they were mere school-boys to old Aristotle." In examining the letter in which Darwin makes this claim and other writings, Gotthelf (1999) suggests that Darwin was most impressed by Aristotle's functional explanations of biological phenomena. Darwin found in Aristotle, however late in life, the "ancient equivalent both of the great modern systematist and of the great modern advocate of comparative functional explanation" (Gotthelf 1999, p. 3).

[10]Lennox (2010) describes a similar movement of the argument in terms of narrative unity. He suggests that Aristotle's *PA* I is analogous to Euclid's *Elements* in that "both [the Elements and *PA* I] are constructed in such a way that the arguments later in the treatise depend, in complex ways, on conclusions established by earlier arguments" (p. 60). Lennox puts it more strongly when he suggests that "like reading a plot, one cannot jump around randomly in the text and hope to fully understand what is going on" (p. 61). I gave some serious thought to organizing my interpretation of the *PA* around topics or themes. There would be a chapter on hearts, a chapter on mouths, etc. In thinking about this means of organization, I realized that pulling together the various, "scattered" discussions of something like the heart did damage to the movement of the argument. For example, a discussion of the heart might appear in a consideration of homogenous parts but also in the context of a discussion of the brain's function in the organismal whole. Organizing the treatment

Aristotle points to the fact that one needs to look at the actions (*ta praxeis*) of animals; he makes a distinction between the actions of animals which are common, those which are according to kind (*genos*) and those which are according to form (*eidos*) (*PA* 645b21–28). Throughout *PA* II–IV a similar distinction is shown within the discussions of *ergon*,[11] which is usually translated as "function" or "work." The last stage Aristotle marks out—examining the actions according to *eidos*—turns out to focus on the function (*ergon*) of the form (*eidos*). An analysis of the function and work (*ergon*) of organic parts and wholes emerges as a crucial ingredient for the study of nature in particular and the study of substantial being in general.[12]

Within the history of biology, the debate about the relative importance of form and function was often quite lively; it is, perhaps, not surprising that the issue of function is first introduced in the *PA* during a reflection on shape (*morphê*). In his analysis of nature, Democritus, according to Aristotle's history, defined each thing according to its shape (*morphê*) (640b34). Aristotle criticizes such a position on the grounds that one could not distinguish between a corpse and a living human, or between a marble hand and a hand of flesh and blood. A corpse is not a human insofar as the former cannot perform the function (*ergon*) that makes a being human (641a4). The appeal to function puts the emphasis on the thing's being alive, on its being in motion.

As we have already noted, Aristotle is interested in looking at actions and movements of animals and animal parts that are common, according to kind (*genos*) and according to form (*eidos*). This breakdown ultimately has a bearing on our understanding of the dialectic between the teleological and the necessary explanations of the cause of certain parts. A teleological account might be thought to require an appeal to a maker of some sort. Looking at the natural, organic world is like looking at the product of some process guided by design. We see an order that we find hard not to ascribe to some conscious maker, some *demiourgos* or a nature that makes nothing in vain (*periergon*). Organisms seem to be made as to be perfectly adapted to their environment. Thus, the flat fishes, things like sole and tongue fish, have, among many other amazing features, their eyes on one side of their head so that they might see better from a concealed position while lying on the sea floor (Fig. 1.9). Flat fishes seem to be designed as if to fill that niche. As Aristotle is eager to point out, the various kinds of birds have various kinds of beaks (Fig. 1.10). Broad, flat beaks are useful for rooting around in marshy areas for food; the curved beaks of raptors are useful in ripping and tearing flesh. These different beaks seem as if they were constructed to fill these different roles. That is not to say that even the most powerful and resourceful craftsman is not constrained; any maker,

by topic undermines the "plot," hampering the attempt to understand what is going on. For this reason, the interpretation follows the unfolding of the text.

[11]e.g. 655b10, 657a4, 658a33, 659a24, 659b30, 661b, 662a17.

[12]Further evidence for the connection can be found in the Metaphysics: "The end is work (*ergon*) and the work is an *energeia*. Hence the term *energeia* is derived from work (*ergon*) and points to *entelechian*" (Meta 1050a22).

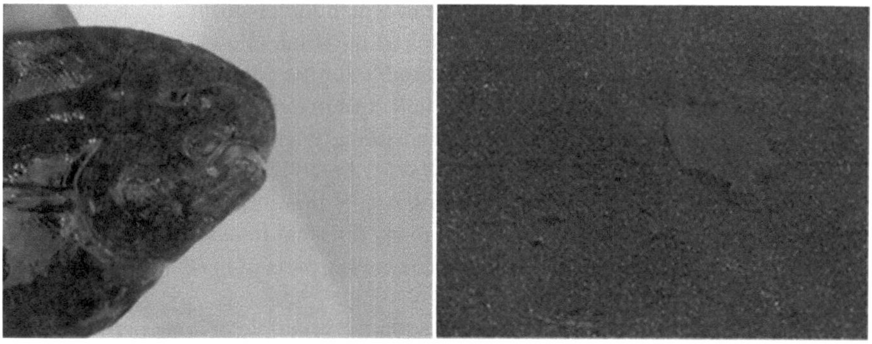

Fig. 1.9 A flat fish concealed on the sea floor

Fig. 1.10 A great white pelican on the Gulf of Gera and a buzzard (*Buteo buteo*)

one realizes on reflection, must be constrained to some extent by material. When one recognizes the necessary, one is already engaged in the attempt to determine causes other than the final cause, which may be the aim of a craftsman of some sort. The wonder experienced upon viewing the seemingly perfectly ordered organic whole is replaced by an attempt to determine the full array of causes behind it.

The tight fit between organism and environment, a fit seemingly produced by a craftsman of some sort, would seem to entail that there be a perfectly ordered relationship between the functions needed and the parts that perform such functions: one part would correspond to one job (683a20–26). However, because of the constraints of necessity—the character of the given material and the harsh necessity of life—the making nature has to construct things differently from what might have been thought best in itself. There is some trade-off between what is best and what is possible within the constraints of the necessary. This results in the recognition of the limit of the principle of one part, one job.

Aristotle often begins by giving the common function of a given part, for example the nose functions commonly as an organ of smell and respiration. He then goes on, in many cases, to show how a part with a common function can be pressed into multiple works, for example the way in which the elephant's nose is used as a hand.[13] This peculiar, specialized function of the trunk allows the elephant to do specialized work. And that in turn is responsible for the continued presence of this part. The elephant's trunk is an organ of smell, aids in respiration, is used as a hand, and, acting as an even more specialized breathing device, allows the elephant to live in watery environments (658b32–659a15). Aristotle's account indicates both the common and the more particular attributes of this part which show how the elephant's trunk is pressed into service in many different functions. The case of the elephant's trunk exemplifies a general pattern: many parts of many animals seem to perform more than one function.

The appealing picture of order expressed in the principle of "one part, one job" becomes more complicated when one looks at the particular organs and actions of organisms. Not only are parts pressed into performing multiple functions, but the study of nature also reveals parts that malfunction or are poorly suited for their function, as we glimpsed above in the brief discussion of cephalopod mouth parts. In other words, when compared to artifacts created by some human design, the parts of organisms leave something to be desired. Our study of those parts leads us to cast doubt on the notion of a making nature. This problem reaches a peak when we are confronted with parts that have no discernible function. For example, Aristotle notes the existence of horns on certain deer that do not perform their usual function; in fact, it appears as if they do not perform any function at all. He even suggests that such horns are at least slightly injurious to the possessor (664a3–8). If nature did nothing in vain, there would be no opportunity to find fault with animals.

The issue of horns comes up in Aristotle's examination of bulls. Why are their horns placed where they are? Aristotle notes that Momus in one of Aesop's fables finds fault with the placement of the horns on bulls:

> Zeus had created a bull, Prometheus a man, and Athena a house, and they selected Momus as a judge of their handiwork. Momus was jealous of their creations and began by saying that Zeus had made a mistake in not putting the bull's eyes on the ends of his horns so that he could see where he was striking, and he said that Prometheus was wrong in not hanging man's heart on the outside that scoundrels could be detected and so that it would be evident what everyone had on his mind. Finally, he said that Athena should have put wheels on her house so that a man could easily move if he had a bad neighbor. Zeus lost his temper with Momus over this spitefulness and threw him out of Olympus. (Daly 1961, pp. 136–137)

By finding fault in the placement of the horns, Momus criticizes the craftsmanship of the gods. But Aristotle is critical of Momus for his particular criticism, not more generally for the act of finding fault with the work of the gods. Finding fault with the parts of animals—as in the case of the horns of deer—may be a necessary refinement of the assumption that nature does nothing in vain. The fact that one can find fault with the arrangement of some parts of some animals must be an important

[13]See Gotthelf's (1997) discussion of the elephant's nose in which he identifies different features that are at work in explaining the extraordinary length and character of it.

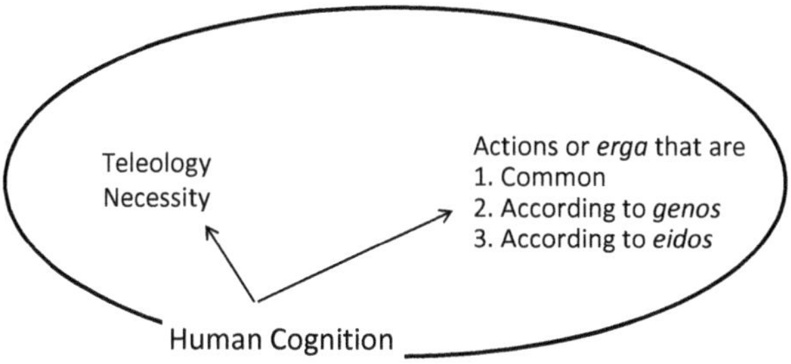

Fig. 1.11 A representation of the way in which humans explore the organic world, which is also a representation of the movement of the *PA*. Necessary functions will, in general, be described as functions or actions which are necessary. Thus, the two foci connect in important ways. This is an important way in which to see the discussion of *PA* I setting up questions and themes that are explored in *PA* II–IV

condition, then, for the inquiry into nature; it provides a starting point for inquiry that would not be available if nature produced a faultless, seamless whole.

The fact that some organs do not adequately perform their function often means that it must be done by other organs, since certain functions must be performed for the integrity of the whole. Many ungulates have multiple stomachs, and one has to understand the cause of such an arrangement. Every one of the horned animals (e.g. sheep, ox, etc.) has several stomachs: Since the workings of the mouth on the nourishment, because of its lack of teeth, is deficient, the stomachs will "receive the nourishment one after another—the first unworked, the next more worked up, the next entirely so, the last finely ground" (674b8–13). The cause of multiple stomachs in some animals is a result of the deficiency of the teeth and mouth: the deficient nature of one part has an effect on the form and function of another part. It has this effect because both the defective or lacking part and those parts that take up the slack are embedded in a whole of parts, so that in this sense the whole is a cause of the partness and organization of the parts.

The necessity of some activities and functions for the maintenance of an organic whole indicates that the necessary cause is not simply material cause. The necessary functions are, in general, what Aristotle describes as the common functions. The issue of the necessary and final causes is thus connected with the consideration of animal function and actions divided into the categories of common, generic, and particular or according to *eidos*. These two issues—the relationship between the teleological and necessary and the way in which actions can be seen as common, according to *genos* or particular—form the foci around which the *PA* moves (see Fig. 1.11).[14]

[14]Kullman (1985) sees a similar structure, calling some functions "secondary" (p. 174). I will develop an interpretation that identifies three levels of function, from the more general to the specific.

What we are as human beings depends in part on certain particular functions of some of our parts. For example, many animals have teeth that are used for the common function of chewing food. However, as Aristotle makes clear, human teeth, while performing the necessary and common functions, serve the more particular function of speech. The issue of human cognition and that distinctively human function or work, speech, will be a thread that becomes manifest at various junctures in the argument of the *PA*. As we shall see, the book opens with some general reflections on the different ways to knowledge. In this way, the actions or functions of certain parts that are according to human *eidos* are reflected in our very analysis of the parts of animals.

In examining and describing attributes or *erga* that are common, according to *genos*, and those that are particular, Aristotle suggests another aspect to the study of causes of organic parts and lives. To appreciate fully the significance of his examination of the range of functions of various parts of various organisms, one must see how *ergon* is identified with what is traditionally understood as formal cause in *PA* I (640b30 ff.). The cause of the maintenance of animal parts through generation is found in their functions. Through its appeal to final causes, necessary causes, and *erga*, Aristotle's *PA* generates a powerful analysis of such fundamental problems in biology as the generation and maintenance of animals and their parts, the relation between organic and inorganic matter, the question of substantial being, and the diversity of life forms. In doing so, it provides a way into more general ontological problems such as the role of function in determining form, the articulation of kinds, the relation between necessity and the good, and, of course, the relation that parts have to the whole in which they are embedded.

Chapter 2
The Problem of Beginnings (*PA* I.1)

Abstract As is often the case in the opening books of Aristotelian treatises, *PA* I is difficult to interpret because of the way in which all of the elements of the whole argument are anticipated. After some general thoughts on the division of knowledge into different kinds, Aristotle touches upon many of the most seemingly intractable problems to be explored in the analysis of the parts of animals. The issue of whether to examine the common attributes of animals or the more peculiar ones is examined (639a15 ff.), but not answered until the end of Book I. Aristotle contemplates whether one should begin an inquiry into nature by taking up the substantial beings (*ousiai*) themselves, such as human beings or lions. The issue of the inquiry into animal life turns to a more general reflection on the relation between phenomena that undergo generation and corruption versus things that do not, phenomena that are eternal; this distinction seems to distinguish zoology, or the study of the animate world more generally, from the study of mathematical physics.

Once we recognize the difference between the sphere that is ruled by generation and corruption, and the one that is eternal, we are able to wonder about the causes of generation (639b11). Phenomena that undergo generation and corruption are ensouled things that have certain motions that are to be examined in our study of organic nature. This is all part of a study of nature that requires an examination of matter, substantial being (*ousia*), moving and final cause (641a26). While it appears as if Aristotle's predecessors do not adequately appreciate the difference between the organic world and the cosmos, Aristotle presents a history of philosophy in order that we might learn from their missteps.

In addition to the more familiar causes in Aristotle (material, formal, efficient and final), there is an extended discussion of the different meanings of necessity (639b20 ff.) that anticipates the claim that cause can be spoken of in terms of two categories, final cause and necessity (642a2). The relation between the teleological and necessary is one of the most important foci of the *PA* as a whole, as we shall see.

J.A. Tipton, *Philosophical Biology in Aristotle's Parts of Animals*, Studies in History and Philosophy of Science 26, DOI 10.1007/978-3-319-01421-0_2,

2.1 The Divisions of Knowledge

The *Parts of Animals* (*PA*) begins with some general comments regarding methodological and epistemological questions. Investigation (*theoria*) and systematic inquiry (*methodos*) are both activities that yield knowledge. Just as there are two **ways** to knowledge, knowledge itself can divide into two: on the one hand we have scientific knowledge (*epistêmê*) and on the other hand we have a sort of educated-judgment (*hoion paideian tina*) (639a5). One would expect the doubleness of knowledge to be a result of the two ways or activities (*theoria and methodos*), but this is left open. If knowledge itself is divided into different kinds, we might ask, from what perspective is this division made? Is it a product of scientific knowledge or of educated judgment or something else? As we shall soon see, it looks as if our understanding of educated judgment can be refined on the basis of a distinction between part and whole. This possibility is suggested by thoughts about parts and wholes that opinion or convention is shown to produce.

> [I]t is a mark of an educated man to be able to judge successfully what is beautifully (*kalos*) stated and what is not beautifully stated by a speaker. Indeed, we think that a man is wholly (*holos*) educated if he is of this sort [about all subjects] and a man has received a well rounded education if he is able to do this. But we consider (*nomizomen*) this man to be unique [one-of-a-kind (*arithmon*)] in his ability to judge almost all subjects, whereas a man may be limited in his judgment only to things of one nature (*phusis*); for there may be an other who is so disposed only to part (*morion*) [of a whole] (639a4–12).[1]

In a book about parts of animals, the first example we get of parts and wholes concerns judgment and knowledge, determined by education. It is, apparently, a matter of convention (*nomos*) that there is something unique about the person who is thought to have a whole education, who can judge what is beautifully said or not beautifully said on a range of subjects. Convention, presumably in contrast with nature, seems to be at work in the difference between educated judgment and scientific understanding; that is not to suggest that there is not something more fundamental that convention points to. The division between two states of knowing is based on a distinction between a concern with a part as opposed to a concern with all subjects, with the whole sphere of human knowing.

The capacity said to constitute a "whole education" sounds as if it is description of rhetoric or some kind of speech-making capacity, although it is unlikely that would be sufficient (Lennox 2001a, p. 120). In contrast to the one who is considered to be wholly educated, there may be those who limit their judgment to things of some nature (*tinos phuseôs*) or to a part (*morion*); it seems likely that this is a refinement of the way in which we are to understand the one with scientific knowledge. While convention has been associated with talk of the whole—we deem

---

[1]Translations of the Greek generally follow Balme (1992a, b) and Lennox (2001). In many ways, Lennox's translation of this opening passage is superior, but the hyper-literal rendering that is offered here is important to highlight certain connections between the opening lines and the treatise as a whole.

someone wholly educated who can speak about all things, or at least judge speech concerned with all things—nature (*phusis*) is associated with the object of partial knowledge. The ability to judge what is beautifully spoken or not on a range of subjects reminds us of Socrates (e.g. *Apology* 21c–22e), although Socrates would never claim to be "wholly educated" (e.g. *Apology* 23a). If it is correct to think of Socrates in the current context, he would begin and end *PA* I.1 (cf. 642a30). Given Aristotle's description of Socrates' turn to the human things, a concern with Socrates, or even with speech and the art of speaking, seems like a strange way to begin an investigation into the parts of animals.

While the one said to have a whole education seems to be held up as superior, nature appears on the side of the one disposed to a part (639a11–13). But if the latter recognized the part as a part in a way the former does not, he would in fact be superior. Aristotle suggests why this is so in the *Metaphysics*:

> There is a science which studies being qua being, and what belongs to it essentially. This science is not any of the so-called 'partial sciences'; for none of those sciences examines universally being qua being, but, cutting off some part of it, each considers the part which it cuts off insofar as it is cut off, as in the case of mathematics (1003a19).

The "partial sciences" do not necessarily recognize the part as part, but rather think that the part they are studying is in fact the whole. Mathematics takes a part but thinks it has a whole. Studying a part and, more importantly, recognizing that you have a partial view might be a way into the study of being qua being. As we shall see, Aristotle's discussion of substantial being (*ousia*) in the *PA* is rich and interesting.

The examination of the parts of animals is explicit in admitting the partial character of its subject matter, so one would think that it escapes the potential defect of the so-called "partial sciences." In the first paragraph of the *PA*, we see issues—division into kinds, part/whole analysis—that will thematically dominate the inquiry that follows. These issues, which will be central to the inquiry into animals, make their first appearance in a consideration of education and states of knowing. The analysis of kinds of knowledge and judgment hinges on the positing of a whole of parts, even if the whole of parts seems to be derivative of conventionally held beliefs about what it means to be "wholly educated."

The split in ways of knowing is overcome or refined when Aristotle exhorts those who are philosophical by nature to experience the pleasure examining cause in ugly animals (645a7 ff.). The person capable of a kind of general educated-ness might find pleasure and benefit from a detailed study of animals.

2.2 Three Problems Concerning Methodology

In reading this book, the first thing we encounter is the assertion that we are not engaged in investigation or systematic inquiry. Rather, we are engaged in *historia*. More precisely, it is *historia peri phusis* (639a12). In the *Phaedo*, Socrates reports his first sailing as an "inquiry into nature" or the "story of nature" (*peri phuseos historia*). Is Aristotle in some way associating his investigation of nature with Socrates' description of his first sailing? Socrates describes his desire to determine the cause of any single thing becoming, passing away, being the way it is. The motivation involves a search for causes, which, for Aristotle, requires an examination of function.

The *historia* concerning nature should have certain rules by which good judgment can be displayed (639a12). Surprisingly, such a judgment does not have as its standard truth and falsehood: "It is clear, then, that in the *historia* concerning nature there should be certain rules by reference to which one shall be able to display good judgment concerning things presented to him, regardless of whether the statements made are true or false" (639a12). This is a remarkable statement, since one might assume that an inquiry into nature would aim at discovering the truth of nature. However, the "good judge," to which this study is supposed to lead, is apparently not immediately concerned with truth or falsehood. Perhaps the statements judged are not conclusions reached in the inquiry. Or maybe the inquiry terminates in questions and not claims to be evaluated as true or false. *Historia* seems to be a part of the educated judgment that is contrasted with more scientific knowledge (*epistêmê*) (639a5) and will be put to use in appraising the method of demonstration. Are we to assume that an examination of animal parts is a *historia*, rather than an *epistêmê*, in being presented as a capacity to judge the speech or method of demonstration? If, in fact, every *epistêmê* assumed certain principles that cannot be examined by that *epistêmê*, it would always require some other capacity, one which could appraise and examine those principles.[2] Those interested and capable need to move toward or beyond the *epistêmê*—either that concerned with animals or some other subject—and engage in examination of the principles of the investigation. The fact that the formulation "*historia peri phusis*" is the way in which Socrates describes his first sailing leads me to suspect that such an activity is not to be associated with the educated judgment described as a capacity to judge speech. Aristotle seems to be implying that his inquiry delves into the territory explored by the pre-Socratic Socrates, but not in a pre-Socratic manner. Any *historia peri phusis* will necessarily be devoted to a study of being by way of an analysis of parts and wholes, as we shall see.

Perhaps some light will be shed on the status of the *historia* and on the general nature of our inquiry by the three examples Aristotle gives of problems about which the one practicing *historia* should be able to make proper judgments.

First Problem—Should one investigate things common to kind or form; or the differentiation (diaphora) *of locomotion* (diaphero)

[2]As Lennox (2011) argues, Aristotle is a "'localist' when it comes to scientific first principles—not just in the sense that each science has first principles peculiar to it but in the sense that those principles will be discovered only by attending to facts that are specific to the domain that they govern" (p. 25).

Should one begin: (a) with the substance, the substantial being (*ousia*), examining separately, for example, the nature of man, a lion, an ox?; or (b) with what is common [to the kind], then proceed to the attributes which are common (*ta koinê sumbebêkota*) [to all the forms] (639a15–17)? Does one start with particular kinds or rather does one begin with attributes common to many kinds of animals? It appears as if Aristotle is questioning whether one can look directly at the substantial beings, the *ousiai*, which Socrates looked at only indirectly through speech (*Phaedo*).[3] Substantial being (*ousia*) here is associated with kinds, like human or lion, as opposed to individuals (cf. 644a24). There are certain characteristics and attributes which cross kind boundaries, such as sleep, respiration, growth, decay, and death, and, in addition, other such passions and dispositions not yet clear (639a20). One cannot use such characteristics to divide up kinds; they are not good at marking boundaries. But an examination of trans-specific attributes is useful for understanding nature. The objection against discussing them form by form (*kata meros*)[4] seems to be a concern with repetition—"we shall often be repeating the same discussion"—since each of the animal forms shares in the common attributes, i.e. growth and death.[5] This problem rests on a split between attributes (*ta sumbebêkota*) and the animals that possess them. Common attributes are characters which occur in different kinds of animals, but which do not themselves allow for differentiation (639a27–29). Can one study attributes without examining the animals in which such attributes are exhibited? Can we look at the *eidê* of attributes without a concern for the actual animals, or do attributes have to be studied only in direct connection with the animals to which they belong? Aristotle appears to be considering the possibility of surveying attributes removed from animals as a way of proceeding.

While focusing on more general properties can potentially be an important way in which we understand particular kinds, an examination of common attributes is not free of difficulties. An interesting difficulty arises when what we think is a common

[3]Lennox (2001, p. 121) notes that "*PA* I makes use of this term [*ousia*] in a variety of ways, and is one of the more interesting texts in which to explore its meaning."

[4]Aristotle uses the formulation "*kata meros*" here (639a 23). The translators take it for granted that this formulation is to be rendered "species by species", implying that *meros* is used as a synonym for species. A more literal translation would be "according to part" which would draw the reader's attention to the problem of the relationship between parts-whole analysis and classification. Lennox (1987, p. 115.) renders the phrase as describing in "partial terms" what belongs universally.

[5]On this, Balme (1992a, p. 73) notes that "Since [Aristotle] takes for granted that the aim of a zoological logos is not merely to describe but to explain, he may also assume that the first necessary step is to pick out correctly the fundamental generic attributes, because they either are, or point to, the causes of the specific attributes: without the generic attributes, explanation cannot begin. Having once stated a generic attribute, one would obviously not want to repeat the explanation for every instance: that would be absurd because it would show that the expositor had not understood the fundamental character of the cause. If this is what was in Aristotle's mind it is admittedly odd that he did not say so. But it could be because he was not ready with evidence. So he leaves it with an apparently trivial reason—the tediousness of repetition—which may be ironical." This is one of the only references that I am aware of in the secondary literature that suggests an ironical aspect to Aristotle! Compare Lennox's (2001, pp. 122–123) discussion of *genos*, *eidos* and *diaphora*.

attribute, possessed by many different animals might have specific differences, which have to be divided further. What is common might not be common when the particularities at the level of form are examined. Aristotle gives animal locomotion as an example of an attribute that can be further divided. Locomotion is thought to be a one, a form or *eidos*. Aristotle uses *diaphero,* literally the "moving-through" of-animals, as the first instance of the need for differentiation (*diaphora*). Recognition of different kinds of motion provides a framework to divide what is thought to be indivisible.[6] The concept of animal locomotion, which seems to be firm and well grounded, is itself put into motion, so that differences emerge.

All the modes of animal locomotion are called "traveling" (*poreia*), but flying, swimming and walking exhibit important differences (*diaphora*); specifically, they do not appear as one in form (*mia to eidei*). What we call by one name—for example, traveling—is upon further reflection divisible into distinct forms or *eidê*. If the nature of nature is to be understood as the source of motion and rest (*Physics* 192b23), it is significant that Aristotle demonstrates the way in which animal motion exhibits particular differences. Other examples of common attributes—sleep, growth, decay and "other passions and dispositions" (639a20)—raise the same question: Can we divide these phenomena that we think are more generic into particulars, into distinguishable forms? Only at the end of the first book (645b1) do we get an answer or judgment concerning this first problem about the starting point of our *historia* (639b6). As we shall see, Aristotle suggests that in our investigation into the organic world, it is preferable to begin with what is common to animals and then move on to the more specific level.

Second Problem—Things generated vs. eternal things; or hypothetical necessity vs. necessity without qualification

The second problem confronting the judgment required in practicing *historia* makes the first explicit reference to parts, to the requirement for knowledge of cause, and to the question of the relation between examining parts and discussing causes (639b10). Aristotle wonders if the physicist who studies nature, like the mathematician, should start first with the observed phenomena (*phainomena*) and parts (*merê*) belonging to each form of animal separately and then state the why, i.e. their causes, or proceed in some other way. The "phenomena" seem to refer to the

[6]Plato's *Cratylus* suggests, however playfully, some connection of *historia* with motion and rest: "Let us first take up again the word *epistêmê* (scientific knowledge) and see how ambiguous it is, seeming to indicate that it makes our soul stand still (*histêsin*) at things, rather than that it is carried round with them, so it is better to speak the beginning of it as we now do than to insert the epsilon and say *epeïstêm*; we should insert an iota rather than an epsilon. Then take *bebaion* (firm), which expresses position and rest, not motion. [437b] And ***historia*** (inquiry) means much the same, that it stops (*histêsin*) the flow. And *piston* (faithful) most certainly means that which stops (*histon*) motion" (437a ff.). This is one of only three occurrences of the word *historia* in Plato—one of the other occurrences is in Socrates' intellectual autobiography and the other is in the *Phaedrus* (244c). Cf. the claim in the *Physics* that the nature of nature is to be understood as the source of motion and rest (192b23ff.).

attributes (*sumbebê*) introduced in the context of the first problem, in which case we might wonder whether the split between attributes and animal is still operative or if looking at the phenomena does not allow for such a split. In any case, the "parts that belong to each kind" brings us to specific kinds of animals. The movement from the first to the second problem seems to be a movement from the common attributes to the specific. The same kind of movement (from common to particular) can be seen in the *PA* with regard to the issue of function.

We might ask ourselves why this would even be a question, since it seems to go without saying that one must start with the *phainomena*—the "what"—before stating the cause—the "why." Perhaps the reason has to do with the fact that it appears as if the postulation of causes does not always come from the *phainomena*; sometimes it comes from the answer to a previous question of why.[7] In the *PA*, it is not always the *phainomena* themselves that move the argument to the question of cause; the postulation of an efficient cause, for example, is a result of the recognition of the inadequacy of material cause as an answer to certain questions, not the result of observing the *phainomena* themselves. It is interesting that in the context of this second problem, mathematics is assumed to begin from the phenomena; but this is qualified in that mathematics is tied to the phenomena in its service to astronomy (639b8).[8] What we might think of as the mathematical abstraction from the phenomena will return in *PA* I (641b11).

Talk of causality issues into talk of genesis: "There are many causes (*aitias*) of generation in nature (*tên genesin tên phusikên*)" (639b11). Of the causes most familiar in Aristotle, the first two we encounter are the final cause and the cause of movement (the efficient cause). Since there are more than one "we should specify which of these is by nature first and which is second. It appears (*phainetai*) that the one called 'that for the sake of which' comes first; for this is the *logos*, and the *logos* is the starting point alike in works of art and in things which are composed by nature" (639b11). As something that "appears" (*phainetai*) first, the postulation of the cause becomes a phenomenon. The final cause is the reason, the *logos* (639b15). In the case of the arts, it is obvious that blueprint as final cause directs the generation of a house or any artifact. In the case of medicine, the doctor looks to health as the end (639b17, cf. 640a29). But Aristotle suggests that the final cause and the beautiful (*kalos*) exist in the works of nature (*en tois tês phuseôs ergois*) more than

[7]This is especially clear in *Metaphysics* 984b8 ff.

[8]This is not always the case. The mathematically inclined Pythagoreans disregarded the phenomena in their positing, based on rational principles, ten planetary bodies when there are only nine. The Pythagoreans played an important role in the history of philosophy presented in Book I of the *Metaphysics*: "they regarded the principles of mathematical objects as the principles of all things" (985b26). They are the first people to draw consequences from a pure argument; they disregarded the phenomena, e.g. "they said that the bodies which travel in the heavens are ten; and since the visible bodies are nine, they added the so-called 'Counter-earth' as the tenth body" (986a11). The Pythagoreans illustrate an understanding of mathematics in which the worldly phenomena are disregarded.

in works of art (639b20).[9] If this is meant literally, final cause is *in* the things of nature as opposed to the blueprint that directs the construction of an artifact.[10] An artisanal product has no internal goal; we supply these craft products with functions and ends.[11] It seems likely that, in the current context, the works of nature, the *erga*, are living beings, yet Aristotle leaves it slightly more general in saying that final cause rests in the works of nature.

After final and efficient causes are introduced, necessity as a cause is put forward (639b20).[12] Hoping to illuminate the necessary, investigators give causal accounts which culminate in the necessary being identified as a cause (642a2). But Aristotle's predecessors did not explicitly refer to the necessary as a cause. The neglect of necessity is based on a neglect of genesis; the predecessors were interested in a *kind* of necessity, but Aristotle intimates that they failed to recognize another kind. "Necessity" turns out to have a number of different meanings.

The conclusion about the final cause and the beautiful being in works of nature is immediately followed by a discussion of the different meanings of the term "necessity." Simple (*hoplos*) necessity applies to things which are eternal (*aidiois*),[13] whereas things existing of necessity by hypothesis (*hypotheseos*) belong to things which are generated (639 b 24–26). The division between the eternal things and the generated things goes together with a corresponding division in what we call "necessity." What is primary, the division in necessity or the division between the eternal and the generated is not clear. Saying that all is water means that all is water eternally. The Pre-Socratics, in replacing mythology with their accounts of nature, might be understood to be replacing the gods with another form of the eternal, by positing some principle that is of a simple necessity.

The division in necessity—which belongs together with the division between the eternal and the changing—illustrates a difference between what might be called physics (*phusikon*) or natural science—I am not sure how best to render the word—and the other theoretical sciences (cf. *Metaphysics* 1025b). The theoretical sciences begin with what already is, while in natural science the starting point is that which

[9]While the beautiful may point to wholes in general, it looks as if it points to the eternal or the perfectly ordered, to the gods. But Aristotle gets to the eternal through a discussion of necessity, as we shall see. See Lennox (2001, pp. 133–134) for a discussion of the ways in which the analogy between artistic and natural production breaks down.

[10]Support for this is found in Aristotle's coinage of the term "*entelechy*." In his essay concerning Aristotle's *entelecheia*, Ritter (1932) suggests that the term be thought of as wholeness rather than purpose. Such a suggestion is very useful in looking at the *PA*, and may be what is at work in the present context. The word occurs in the first chapter of the *PA* at 642a in the talk of animal generation.

[11]Broadie (1990) discusses Aristotle's use of the "craft analogy."

[12]As Lennox (2001, p. 126) notes, "it is a philosophical and scientific issue what sorts of things can be causes, an issue as hotly debated in the ancient world as in the modern." (Cf. Frede 1980; Freeland 1991).

[13]cf. *Generation and Corruption* 337b14.

will be (640a4).[14] Physiology differs from the theoretical sciences by examining another form of necessity. Thus, it deals with the generation of the non-eternal.[15]

When asking about generation in nature, final cause and moving cause are two among several (639b10). But those giving a causal account hope to refer their explanations to necessity. But as we shall see, understanding the necessity involved in generated things allows one to tease out another cause, the material (639b25–27). The matter in generated things will of necessity pre-exist. The question becomes whether the sequence of generation will also be of necessity. We can say that in contrast to the *phainomena* that exist necessarily without qualification, which are the objects of the so-called theoretical sciences, the *phainomena* studied by physics or natural science exist by hypothetical necessity, as things undergoing generation and corruption.[16] This observation forms the transition to the third problem.

Third Problem—Leaving mathematics and the eternal behind.

In what way does each thing come to be generated by nature? This is the third problem put before the one judging the principles of an inquiry into nature (640a11). It is not immediately clear how this problem is significantly different from the issues entailed in discussing the second problem. We can, however, start by saying that the second problem emphasized the causes of generation, but more importantly it pitted generation against that which is eternal (639b24). With the third problem, the concern with the eternal has been left behind. The issue now is the difference between how each thing is generated and the way in which it exists. Each generated thing, which would include every organism, is a substantial being (*ousia*) that is the product of a process of generation. In this way, *ousia* can be seen as the "for the sake of which" of genesis.[17]

Within the context of the third problem, Aristotle discusses animal morphogenesis. An animal develops with a backbone because it is the offspring of something with a backbone. Aristotle accuses Empedocles of not understanding the necessity at work in generation (640a20). Empedocles thought that the backbone

[14]See Lennox (2001, pp. 129–30) for a discussion of the way in which the scientific demonstrations of the necessary governing eternal objects and generated ones differ.

[15]The first principle, the material principle, is supposed to be that which does not vary, all is water eternally. Aristotle seems to be showing how physiology or the study of nature ultimately concerns genesis and in this way shows that he is doing something distinct from pre-Socratics. It is not clear whether the contrast we are examining is between the natural sciences and the theoretical ones or whether we are meant to see the difference between these sciences and the crafts. Lennox (2001, pp. 128–130) outlines the various positions defended in the scholarly literature. Like Lennox (2001, p. 129), I see the argument as an examination of the contrast between the natural and theoretical sciences, which is informed and refined by an understanding of generation and necessity.

[16]On this, Dudley (2012) claims that Aristotle "is clearly contrasting physics and the (other) theoretical sciences in regard to the mode of demonstration and the type of necessity found in them" (p. 122). Questions regarding types of necessity seem to multiply at turns in the *PA*.

[17]Such a notion is at work in Plato's *Philebus* when Socrates examines a *genesis eis ousia* (26d). The limit imposed on *genesis* is *ousia*. But unlike the arts where one can point to a table, it is difficult to pick out any particular moment when an animal has reached the limit of genesis.

was by accident, and Aristotle appears to reject this in pointing to necessity in the development. But it is doubtful that Empedocles was arguing that a particular animal develops a backbone by chance; most likely he meant that the characteristic of having a backbone arose by chance.[18] Aristotle argues against Empedocles' reliance on chance by appealing to animal reproduction (640a23). But he does not stop with the formulation that the thing begotten has the same form as the begetter. Not only does the mule (the begotten) have a different form from either the horse or the ass (the begetters) (641b35), but it is sterile. There is no doubt that Aristotle understands the role of chance in nature despite his criticism of Empedocles in this context: "Things produced by art may result in a similar manner also by chance; for they are produced as they would be by art, e.g. health" (640a 33). Health is one thing that can arise by chance, and it seems plausible that Aristotle would entertain the possibility of other biological phenomena arising by chance.[19]

2.3 The Phainomena

After going through the three problems concerning methodology—that of proceeding by common attributes or particulars, that of the relationship between hypothetical necessity and necessity without qualification,[20] and that of the relationship between parts, cause and generation—Aristotle presents a description of the *phainomena*, the beginning point of our inquiry into nature. The discussion of the three problems has given us a broad indication of the elements of our inquiry into

[18]Dudley (2012) analyzes the Aristotelian notion of chance from many different angles. Thinkers such as Empedocles and Anaxagoras—the "Presocratic universal determinists" as Dudley calls them—"saw no contradiction in holding that a necessary event occurred by chance. Thus for them chance appears to refer to the subjectively unexpected nature of certain necessary events" (p. 3).

[19]It might be important to note that this is the second case in which Aristotle appeals to health, the first being at 639b17. Health, as the *logos*, can be the product of a doctor with the art of medicine, or it can be the product of chance. In the former case we have, in a certain sense, a design (health) with a designer, a doctor. In the latter, we do not have the designer or agent, unless chance is appropriately spoken of as playing this role. Or to use the terminology here, the *logos* health can have a creator or not. Alternatively, the *technê* is regarded as the "logos of the work (*ergou*) without the matter (*hulê*)" (640a32–34), and so too should we recognize chance as the logos of the work without the matter. In this way, things like health are the work (*to ergon*).

[20]Cooper (1985) has an extended discussion of what he calls Democritean material necessity and hypothetical necessity in the context of the *Physics*. One of Cooper's aims is to show that Aristotle "holds, of course, that material necessity operates wherever matter is found, but his claim (as I have interpreted him) that it operates only against the background of hypothetical necessity is limited to the formation and behavior of living things. If, for example, ice forms on a pond as a result of material necessities attaching to the natures of cold air and water, nothing he says in these passages commits him to seek some hypothetical necessity to explain why the air and water in question become conjoined, with that result" (p. 163).

nature: (1). *phainomena;* (2). cause; (3) parts and genesis (640a15), with genesis being an additional layer not found in the earlier description at 639b7. The earliest thinkers had analyzed things in terms of cause. Aristotle is suggesting that an exploration in terms of genesis is somehow to go beyond the pre-Socratics. They gave causal accounts thinking they were talking about genesis, and these causal accounts are rooted in an attempt to explain some *phainomena*. But what exactly were the *phainomena* they were trying to describe? Were the phenomena primarily animate?

The *phainomena* we wish to explain in the *PA* are, obviously, natural beings understood as animate wholes consisting of parts. Aristotle uses the human being as an example:

> Hence it would be best to say that, since this is what it is to be a human being, on account of this it has these things; for it cannot be without these parts. If one cannot say this, one should say the next best thing, i.e. either that in general it cannot be otherwise, or that at least it is good thus. And these things follow. And since it is such, its generation necessarily happens in this way and is such as it is. (This is why this part comes to be first, then that one.) And in like manner one should speak in precisely this way about all of the things constituted by nature (640a 33–640b3).

Perhaps we are to understand the *genesis* of humans as the underlying problem of the work. The genesis of the human being is introduced, in any case, to help us understand how the parts of an organic whole are generated, and thus how all things of nature (*pantôn tôn phusei*) are constructed. Those initially interested in the study of nature were struck by the "phenomenon" of the beautiful wholeness of humans. There is a progression from the beauty of the human whole to a recognition of the parts of the whole, then to the *genesis* of one part followed by the *genesis* of another part. Some understanding of the possibility of human wholeness seems to be the way the study of nature begins. Knowledge and articulation of cause gives way to an exploration of the genesis of parts, which has as its motivation the attempt to understand human beings. The discussion of the problems, which move from *phainomena* to cause to *genesis*, culminates in the unveiling of the recognition of the wholeness of human beings. A human being, perhaps more than any other being, confronts another human as a whole of parts.[21] That this is the peak of this particular argument is indicated also by Aristotle's turn to a discussion of his predecessors. By his construction and presentation of a certain history of philosophy, he will presumably discover those limitations which his own investigation will have to overcome.

[21]Of course this is to abstract from the fact that a particular human being does not have all of the parts that compose the human animal, only those of its particular sex.

2.4 The Ancients and Their Causal Accounts: Material Cause

The material principle is said to have been the only cause recognized by the ancient philosophers (640b5). In Aristotle's own opening account, the first causes discussed were the final and the cause of motion (639b12), followed by a discussion of necessity as a cause (639b22), and ending with what looked to be a note on material cause (639b27). In all of this, Aristotle is silent on the formal cause. In the turn to the discussion of his predecessors, not only does the material cause emerge as first, but also the opportunity will arise for a discussion of formal cause.[22] Aristotle's predecessors pondered how the whole is generated (*ginetai to holon*) and by what mover (640b8). They are most interested in the generation of the cosmos (*kosmôn gennôsin*), the cosmic whole, which they hope to explain by appeal to a material principle. It appears as if the predecessors that Aristotle has in mind are not focused on human wholeness. Aristotle indicates that the ancients are necessarily concerned with the generation of wholes—the cosmos primarily—even though they appear to concentrate only on the material principle. If one is to say that all is water, one has to account for non-water like wholes and other phenomena. To posit a material principle is to claim that everything is made out of a particular substance and returns to that particular substance—from dust to dust. Such a structure—the formation from the substance and the return to the substance—implies a sort of completion, a process that can be viewed as a whole, as Aristotle seems to indicate (640b8). He posits that his predecessors, putting forward a material principle, are compelled to face the question of generation. The question of efficient causality is latent in that of material causality.[23] They assume that the underlying matter must have a nature that can be explained by an efficient cause (640b9), whether that cause is strife, mind, or some accident. In Aristotle's account, it is not clear whether the predecessors were trying to explain the nature of the material principle or revealing—perhaps despite their intentions—that they will not be able to explain the phenomena by matter alone, without some kind of source of motion. What does seem clear is that what we think of as efficient cause is described as an underlying nature of material cause in this argument.

As we have already seen in Aristotle's opening account of the problems confronting those conducting a *historia* of nature, prior to the turn to the predecessors,

[22]The closest we get to a discussion of formal cause is the description of the *erga*, or works, of nature and art (639b21, 640a33); I say this in reference to the way in which *ergon* will emerge as a crucial characteristic of forms (below 641a2).

[23]In her interesting treatment of what she identifies as secondary teleology, Leunissen (2010) notes that, "Aristotle does not deny that natural, materially necessitated processes can have beneficial outcomes. He only denies that they can have such beneficial outcomes on a regular basis without the intervention of goal-directed efficient causes, which (actively) adapt or (passively) co-opt such features in order to support the animal's well-being" (pp. 34–35). This will be clear in the discussion of certain residues in cephalopods that are put to a good use.

Fire—material cause

Hot—underlying nature out of necessity (efficient cause)

Fig. 2.1 On this model, efficient cause is understood as underlying material cause, as latent within the matter

the recognition of material cause results from an appeal to necessity after beginning with final and efficient causality (639b12–27). In the Pre-Socratic framework, according to Aristotle, necessity moves these thinkers from material cause to something like efficient causality. Among the string of questions posed by the predecessors include "what and what kind of thing is matter?, how does the universe come to be out of it?, and with what cause of movement (such as strife or love or mind or spontaneity), assuming that the underlying matter has a certain kind of nature out of necessity—fire a hot nature and earth a cold one, the former light and the latter heavy?" (640b6–11) (see Fig. 2.1). This picture, we are told, is how the predecessors generate a model of the cosmos; but they apply this model also to the generation of plants and animals (640b12).

Restricting explanations to material cause leads to a failure to recognize potential differences in the generation of animate and inanimate wholes. One indication that the generation of things in the animate world differs fundamentally from that of the inanimate is to be discovered in the notion of capacity (*dunamis*): the parts of an organic whole can be analyzed in terms of *dunamis* (640b22), which must be added to the materialistic account.

"Most of those who first philosophized," Aristotle remarks in the *Metaphysics*, "regarded the material kinds of principles as the principles of all things; for that of which things consist, and the first from which things come to be and into which they are finally resolved after destruction . . . this they say is the element in the principle of things; and because of this they think that nothing is generated and nothing perishes, since such a nature is always preserved" (983b7). These thinkers in positing an eternal principle also posit something that is necessary without qualification (639b24; cf. *Physics* 191b12). In this account, Aristotle argues explicitly that the positing of the material principle denies, in an unqualified sense, generation and destruction: instead we have aggregation and separation (*Metaphysics* 984a9). The *PA* account of Aristotle's predecessors highlights a different point; emphasis on a material principle immediately points to generation and an efficient cause.[24]

[24]"Thales' 'water' is not really water," Michael Davis (1999, p. 55) claims in his examination of *Metaphysics A*, "it is a thought thing, not a perceived thing. In a way it is not really material at all. Still, Thales somehow knows that the stuff that is to make everything else knowable to us must itself be known. So, although he cannot explain why the hidden true nature of things should sometimes be visible in things (and so unhidden), he does sense that unless this is the case he will not have succeeded in making the world knowable. In addition, Aristotle adds that perhaps Thales assumed water to be the first principle because of its connection to growth . . . . But if water as a

2.5 Efficient Cause

As indicated, in Aristotle's account of his predecessors in the *PA*, the question of efficient cause comes up with the positing of material cause: these thinkers, following the model used in analyzing the generation of the cosmos, say that "as water flowed into the body a stomach and every part that receives nourishment and residue came to be; and as the breath passed through, the nostrils burst open" (640b14–16). In this description, it is hard to say whether air and water are to be thought of as material cause or efficient cause. The very nature of the material seems to play the role of efficient cause. The way in which Aristotle's predecessors understood the relation between material and efficient cause is presented here in different terms from the account given in the *Metaphysics:*

> From what has been said one might think that the only cause is the kind which is called 'material.' But as philosophers progressed in this manner, the facts themselves opened the way for them and contributed in forcing them to make further inquiries. For no matter how much every generation and destruction proceeds from some one principle, or even more than one, why does this happen and what is the cause? For, indeed, the underlying material itself does not cause itself to change. What I mean, for example, is this: neither the wood nor the bronze causes itself to change; the wood does not make a bed, nor the bronze a statue, but some other thing is the cause of the change. Now to seek this is to seek another principle, namely, as we might say, the source which begins motion (*Meta*. 984a17).

Certain realities forced these thinkers to search beyond material cause. In the *PA* account, the Pre-Socratics, in their investigation of the whole (*to holon*), tried to find out what set in motion the process of the formation of a whole; they assumed there to be some underlying nature (e.g. strife, friendship, mind, etc.) of the matter (640b10). According to Aristotle's account in the *PA*, his predecessors examined final and efficient cause for the *kosmos* (640b12), which they considered the primary phenomenal whole, and perhaps the primary, paradigmatic whole. Aristotle's predecessors took what they learned of the cosmos and applied it to plants and animals. Making no distinction between what they viewed as the whole of the *kosmos* and animate things, they failed to recognize the problem of the difference between the animate and the inanimate. That problem will consume much of the discussion of *PA* II, especially II.1–9, indicating that Aristotle did not assume, as his predecessors did, that the explanations given for the inanimate were adequate or appropriate for the animate; one cannot simply apply what his learned about the cosmos to plants and animals.

---

first principle has within it growth—a hidden cause of change or motion—then Thales' material cause is, without his realizing it, also an efficient cause."

2.6 Ergon and Formal Cause

In the course of presenting the thought of his predecessors, Aristotle introduces the idea of function (*ergon*). It comes up in the context of examining Democritus' view that man is defined by shape and color (640b29) or just by shape (640b34). A corpse has the same shape, the same *morphê*, Aristotle argues, but is not a human being. A bronze or wooden hand is not a hand because it cannot perform its activity or function (*ergon*). A dead man is not a man because he cannot perform the function that makes a human being a true human being (641a4).[25] Democritus recognized some formal cause above material cause, but Aristotle criticizes him for overemphasizing shape. Aristotle himself acknowledges that the nature associated with shape is more important than the nature with respect to matter (640b27); but function, it seems, is a way to de-emphasize shape when talking about form—it is more than shape. With the discussion of form as function, we are pushed to ask what this implies about the relation between formal and final cause. As the argument progresses, Aristotle will suggest a way in which actions or movements can be polymorphic (646b15, also 656a4; see Tipton (2011).

As epitomized by Democritus, if Aristotle's predecessors were concerned with form at all, it would only be form construed as shape.[26] They were then pushed, unaware according to Aristotle, to the question regarding the generation of things in their shape (641a9). Aristotle points out something that was implicit in the account of at least some of his predecessors. The problem of *genesis* lay hidden behind the need to determine the cause of form understood as shape: "by what power (*dunameon*) were these things and shapes fashioned (*edemiourgethesa*)?" Aristotle uses a term from the arts to raise the issue of the cause of shape. But in doing so, he raises the question of how exactly the equivalent in nature is to be understood.

The position taken by Democritus is an advance on the appeal to material alone in describing the genesis of the cosmos (640b12). Democritus seems explicitly concerned with animals (640b29), including humans (640b31). In addition, Democritus and those following his lead might ask by what power (*dunameôn*) the parts are fashioned (641a9), thus avoiding Aristotle's criticism of those predecessors who neglected the capacity (*dunamis*) of organic parts (640b19–b24). The *dunamis* of the parts (640b22) is reformulated as part of the question, by what *dunamis* are these fashioned? (641a7), which, on the model of the arts, switches the emphasis from the parts of the whole to some maker of this whole. In this presentation, Aristotle notes that introducing the arts to understand the generation of organic things necessarily

[25]According to Aristotle, Democritus relied on shape and color. In recognizing color, does Democritus attempt to add a dimension to shape? A corpse does not have the same color. And by dropping a concern with color (640b34), is Aristotle's argument against Democritus able to progress more smoothly?

[26]Balme (1970) notes that Aristotle "criticizes the emphasis on morphology, which he holds subordinate to function." (p. 262). See also Balme 1987b, pp. 78–79.

pushes us to final cause: the carpenter will explain why he made such a stroke, or for the sake of what he made the parts of the artifact (641a13).

Aristotle has presented a history of the thinking of his predecessors that places material cause first (640b5), followed by the cause of movement (640b8), to which Democritus adds a concern with form (640b32). Finally, Aristotle points to the fact that final cause is implied when one fleshes out the unstated model of the arts that seems to be behind all or some of the thinking of the predecessors (641a7–14; cf. Lennox 2001, p. 136).

What causes an animal to be the animal that it is?[27] It cannot be form alone, as morphology or shape, that makes an animal what it is, or Democritus would be right and the distinction between bronze animal and live animal would be hard to maintain. A modern example of the shortcoming of placing too much emphasis on form is cryptic speciation—two organisms can have the same form yet live very different lives.[28] The animal does not live as it does by virtue of its shape and its color alone. One must appeal to function, which can be understood as life history when considering organismal wholes (Tipton 2006). One must attempt to understand how the animal lives in the world. Understanding the function of an organism to be life history requires examining the being in its life, in its movements as an ensouled thing.

The discussion of Democritus and his understanding of form leads to a discussion of the question of soul. The issue of form points both to the shape of a thing and to the soul of an animate being, a substantial being. Form can be understood as something that distinguishes from the outside or from the inside.[29] From the outside,

[27]Perhaps we are also pushed to the question, why and how do animals differ at all? It has been suggested (Balme 1987d, p. 301) that Aristotle's answer to this question "is the double explanation, 'necessity' and 'the better'. Given the necessary limitations of heat and environment, each animal form is the best possible: that is, the form which brings it the most functional advantage, what Aristotle often calls 'the useful.'" The dialectical relationship between the necessary and the useful can be seen working itself out in what Aristotle would identify as the *ergon* of organic wholes, their movement and life history. This is something which I explored (Tipton 2006) in relation to Aristotle's thoughts on two different fishes which are morphologically very similar.

[28]Gotthelf (1987, p. 181, n.40) suggests that "identifying what an organ does is not sufficient to explain its presence. One must go on to show why the organisms which have it *need* something that does what it does (or are otherwise better off for having it rather than not). It seems too that for Aristotle to call what some organ does its *function* (*ergon*) is for him to say both that it does that, and that it is necessary (or better) for the organism that it do that". In my opinion, there is much in Gotthelf's tantalizing suggestion that needs to be explored. Gotthelf also suggests how this might be a point of contact with contemporary debates, namely those surrounding etiological theories of function. In addition to expanding on this idea of function in relation to the parts of the animal, I will examine the way in which function applies to the organism and its behavior *as a whole*—*ergon* as the "work" of the organismal whole. This is a departure from Gotthelf and entails examining what it means for Aristotle to refer to the body as an *organon*.

[29]For a detailed discussion of the way in which form-matter relations require an appeal to function see Cynthia Freeland (1987). Robert Bolton (1987) points out that the account of a thing's function is to provide basic theoretical information about it. For suggestions on how function may be understood on the organismal level, see Furth (1987, pp. 26–27, p. 29, and esp. p. 39).

form means shape; but from the inside, it must point to the soul of a living thing. In posing the question, "By what power (*dunameon*) were these things and shapes fashioned (*edêmiougethesa*)?" (641a9), Aristotle's "Democritean" predecessors push the argument in the direction of efficient cause.[30] On the model of the arts, which Aristotle suggests lies behind the thinking of his predecessors; the efficient cause is the craftsman who is altogether distinct from the material. However, the most direct answer to the question of what is responsible for the generation of the given artifact is a tool (*organon*; 641a13). Still, to say "a tool" is not enough; it is the artist with the aid of a tool that we point to when we ask what brought an artifact into being. The efficient cause, the artist with tools, is completely separate from the material (cf. Lennox 2001, p. 134).

On the craft model, there is no problem in distinguishing efficient and material cause. Nature, however, presents difficulties. In their investigation of nature, the early philosophers say that the thing with its shape is formed by air or by earth (641a10). Material and efficient cause are not easily distinguishable in organic things. Hence, it is understandable that an appeal to material cause is made when we expect an explanation in terms of efficient cause. But this explanation has a certain truth, since it does not make the efficient cause something outside and separate, like the artisan and his tool (*organon*). The clear distinction between material and efficient cause on the model of the arts cannot work as an account of the animate things. If the efficient cause were outside, it would not be a part of the thing generated. An artisan is obviously not a part of artifact produced. To apply the notion of efficient causality to organic things requires the transformation of the tools (*organon*) employed in the arts into the organs of living things must occur.

2.7 Soul, Ergon and Nature

When function (*ergon*) is first introduced (641a2) it serves primarily as a means to correct Democritus' emphasis on form as shape. Form or shape is the same regardless of whether the thing is alive or dead. But something dead or inorganic cannot perform the *ergon* proper to its living counterpart: a marble hand cannot grasp. The implication is that a part ceases to be a part if unable to perform the function proper to it; an *ergon*-less part is only a homonymous part (641a). It is in this capacity that soul is introduced (641a19): "When the soul departs, what is left is no longer animal and no part of the animal remains the same, except in shape only, like the animals which turned into stone in the fable" (641a19). Soul is thus identified with an *ergon* of the animal as a whole. It is the job of the natural scientist to study soul because it is the soul, or at least part of the soul, which determines kinds (*eidê*) (641a23). If soul were the *ergon* of the animal, then *ergon* would determine

[30]See Chalmers (2009, pp. 19–41) for a more extended examination of Democritean atomism in the context of the history of science.

Fig. 2.2 A representation of the way in which nature is said to be related to substantial being and the causes

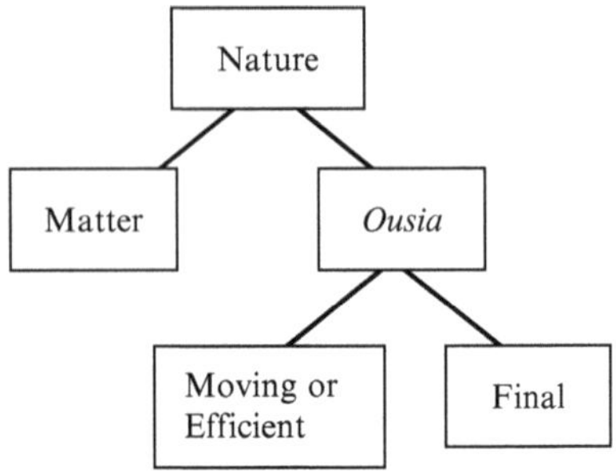

kind. If we understand *ergon* as activity, we can use it to demarcate species like differences in animal travel.

The natural scientist studies soul as a study of nature. We get to the soul as an object of study by Aristotle's correction, via an appeal to *ergon*, of the Democritean emphasis on form as shape. The study of nature prior to Aristotle characterized by the emphasis on causal accounts unknowingly leads up to the study of soul. What Aristotle is keen to point out is that an interest in cause also seems to point to an interest in substantial being (*ousia*), which is something that distinguishes Aristotle's study of nature from that of his predecessors.

The Pre-Socratics' neglect of soul is due in part to their conception of nature. According to Aristotle, nature is spoken of in two ways (641a 26): (1) as matter (*hulê*); and (2) as beingness or substantial being (*ousia*). Thus, the conception of nature is double. *Ousia*, Aristotle suggests, also has a doubleness, insofar as it encompasses both moving cause and final cause: "The term 'nature' is spoken of correctly in two ways, (1) meaning matter, and (2) meaning substantial being (*ousia*), which includes both the moving and final cause" (641a28). The two ways in which nature is conceived point to at least three of the four causes—material, efficient and final (see Fig. 2.2).

The only thing that appears to be missing is form as cause, unless substance or *ousia* set against matter is formal cause; this would imply that formal cause in some way serves as moving and final cause. The result of this identification of formal cause with *ousia* is that nature can be spoken of correctly in two ways, in the case of an animal, as body and soul.

Either the whole soul or a part of the soul is the *ousia*. The opinions about nature point in two directions: away from soul, toward matter or body, and toward soul or beingness, away from matter or body. These three causes—matter and *ousia* constituted by moving and final cause—point to body and soul. He who investigates nature should investigate the soul more than matter, since the latter exists for the sake of the former (641a30). Matter and the parts of the body exist for the sake of the soul, for the sake of living, for the function of the organism. At the end of *PA* I this "for the sake of" will be cast in terms of action (*praxis*) (645b15). Likewise, a body is said to be an *organon* existing for a certain purpose (642a12); this purpose is soul (641a31) and, presumably, the activity of soul. It is in an organism's living or activity that soul is understood as final cause.

2.8 Intellect and Motion

The issue of parts and wholes, which first appeared in regard to education (639a5), is now applied to the entire soul and to the part of the soul that includes the intellect:

> In view of what was said just now, one might puzzle over whether it is up to natural science to speak about all soul, or some part, since if it speaks about all, no philosophy is left besides natural science. This is because reason is of the objects of reason, that natural science would be knowledge about everything... However, it is not the case that all soul is an origin of change, nor all its parts; rather, of growth the origin is the part which is present even in plants, of alteration the perceptive part, and of locomotion some other part, and not the rational; for locomotion is present in other animals too, but thought in none. So it is clear that one should not speak of all soul; for not all of the soul is a nature, but some part of it, one part or even more (641a33–641b9).

As natural scientists, it must be determined whether to study the whole soul or simply a part of the soul, that part having to do with motion. The double conception of nature—nature as matter and as *ousia*, which includes the moving and the final causes (641a27)—leads us to focus on the source of motion (641b5); we are exploring the question of soul and the study of nature by focusing on the realm of movement (cf. *Physics* 192b23). *Physikê* is the study of bodies in motion; if there is a part of the soul that is not involved in motion, natural science is not interested in it as an object of study. Natural science (*phusikê*), then, should not have intellect as an object of its study, if *nous* is not involved in motion. If it had a role in the type of motion being examined, there would be, Aristotle argues, no room for any other kind of inquiry. We are looking for the principle of motion (*kinêseôs*) or locomotion (*phoras*) in a part of the soul. It might turn out that like animal travel (639b2), the motion that characterizes soul has specific differences. Soul motion is to be differentiated.

Plant soul, as the principle of growth (*auxêsis*), differs from animal soul, formulated as the source or principle of alteration (*alloiôseôs*). Human soul, of course, is distinguished by the presence of mind, *noêtikon*. Soul is a whole of parts, and these parts are responsible for different species or forms of motion. Compare this to what Aristotle says about animal locomotion when he divides the general kind "traveling" into many species (639b1). We are looking for the source of locomotion (*phoras*), but that source does not lie in the thinking part of the soul because locomotion exists in other animals without thought. Aristotle concludes from this that the study of the animated world does not concern the whole soul because not every part of the soul is nature (641b10). At the core of nature in the comprehensive sense lies the principle of motion. There are "modes" of locomotion specific to different kinds: animals differ from plants in having the capacity for alteration as opposed to the growth of plants. Plants and animals are distinguished as kinds based on a motion. Natural science (*phusikên epistêmen*) as such does not appear to investigate the motion of thought. At this point we are forced to ask a question regarding the relation between philosophy and natural science. If the former is understood as a study of the whole soul, how does the natural scientist's study of a part of the soul contribute to a philosophical understanding? The short answer is that there is immense pleasure for those who are naturally philosophic in studying animals (645a10). But there also must be something important about getting one's hands dirty, so to speak. When we turn to a study of the non-human animal world, we learn something about teleology, function, chance and the different ways in which animals appear to be fashioned to fit their particular environments.

2.9 Nature Does Everything for a Purpose

The first occurrence (641b11) of the formulation, "nature does everything for a purpose" (*ê phusis heneka tou poiei panta*), is used against the notion that objects of abstraction studied by mathematics should be part of the purview of the study of nature.

> Further, none of the abstract objects can be objects of natural study (*tên phusikên theôrêtikê*) since nature does everything for the sake of something (*ê phusis heneka tou poiei panta*); for it appears that, just as art is present in works of art, so in the things themselves (*en autois tois pragmasin*) there is a kind of source and cause such as heat and cold which comes to us from the entire universe (*tou pantos*) (641b11–16).

The formulation "nature does everything for a purpose" is part of a larger discussion of necessity and chance:

> This is why it is more likely that the heaven (*ouranos*) has been brought into being by such a cause—if it has come to be—and is due to such a cause, than that the mortal animals have been. Certainly the ordered and definite are far more apparent in the heavens than around us, while the fluctuating and random are more apparent in the mortal sphere. Yet some people say that each of the animals is and came to be by nature, while the heaven, in which there is not the slightest appearance of chance and disorder, was constituted in that way by chance and the spontaneous. We say 'this is for the sake of that' whenever there appears to be some end towards which the change proceeds if nothing impedes it. So it is apparent that there is something of this sort, which is precisely what we call a nature (641b17–b28).

The efficient cause in the model of the arts is the artisan; but it is only in a metaphorical sense that we would say that the artisan is in the artifact. Aristotle suggests that the purpose in creating the artifact is transferred in some sense from the design the artisan has in mind into the artifact and that is how the art is in the work of art. The craft product has a purpose only in relation to the artisan and the user of the artifact. Interpreted in this way, the universe would supply us with the hot and cold in order to accomplish some purpose in the same way as an artisan suffuses something with function. However, it is easier to think of hot and cold as efficient causes themselves.

In this context, *ouranos* is set against mortal things, with the order and definiteness of the former contrasted with chance and luck of the latter (641b20). A particular concern is the way in which these different spheres relate to different motions. The order and definiteness of the heavens, for some thinkers, are by nature (641b22); this seems to be the nature that does everything for a purpose. In this conception, the chance, disorder or luck of the mortal sphere would be anti-natural.

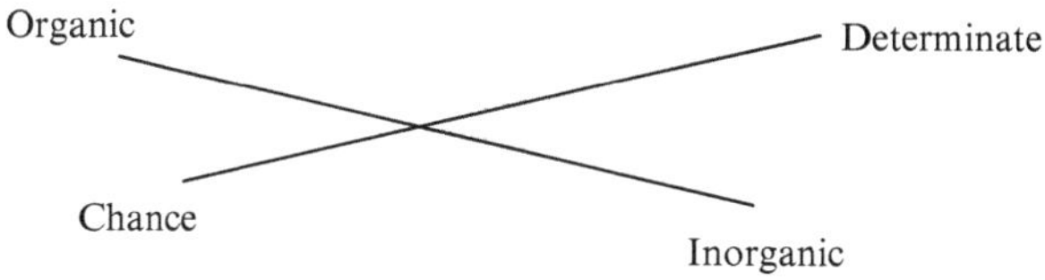

Fig. 2.3 The intersection of two continua. The continuum between the organic and inorganic will be scrutinized in *PA* II

We must determine, however, whether Aristotle views chance as something that is opposed to nature. He notes that whenever there appears to be some end toward which motion, if not obstructed, is directed and finally terminates, we say that this is for the sake of that. This is what we identify by "nature" (641b24). The emphasis here is on appearances (*phainetai*) and how things are spoken of. Aristotle is not yet ready to confirm this view of nature. The division between ordered and disordered was associated with at least some of the predecessors with the division between the inanimate (planetary bodies and the heavens in general) and the animate. The question becomes, does the appeal to the ordered and definite versus the disordered and that infected with chance allow us to understand the distinction between the animate and inanimate?

If chance and luck lie at one pole and order and definiteness at the other, where do we put mortal things on this continuum? Actually, the argument presupposes some continuum between the organic and the inorganic, so that there would be some point of intersection of these two continua (see Fig. 2.3). The assumption that there is a continuum between animate and inanimate will be further scrutinized in *PA* II.

Aristotle, in the previous stage of the argument, suggests that chance and luck exist in mortal things (641b21) as a way to argue against the position that the heavens come to be by chance (641b23). Now the argument turns in the opposite direction. To demonstrate that organic nature, as opposed to eternal things like the planets, is not ruled by chance, Aristotle turns to the generation of animals: "the seed is an origin and is productive of what comes from it. For these things are by nature; at least they grow from seed. But prior even to this is what the seed is a seed of; for while the seed is becoming, the end is being (*ousia*)" (641b27). Generation in the arts is not dependent on chance or luck; an artisan sets out to generate a table and generates a table; this is the making (*poetike*) of the arts. The trajectory of such making is predictable and straightforward. Initially, the generation of organisms also appears to have a straightforward trajectory and Aristotle says that he means to use animal generation to demonstrate the order and definiteness of the generation of the animate. The seed develops into an offspring that is the same in kind to the parent, so that the generation is predictable, not a product of chance or luck. A seed can be viewed from its source or from what it finally becomes. In most cases these two views coincide. However, Aristotle does realize that seeds can sometimes develop into individuals quite unlike the parents, such as asses (641b35). Hybridization is only one example of the unexpected resulting from animal generation. Animals reproduce for their own good. The individual achieves immortality only through reproduction (GA 731b35). Individuals reproduce in a process that preserves their own form. Hybridization is obviously not a preservation of form. Also, sterile hybrids cannot participate in this attempt to acquire immortality through sexual reproduction. The appeal to animal generation in order to support the view of the orderliness and definiteness of mortal things has to be combined with that argument which suggested that chance and luck were present more in mortal things than in the heavens (641b21).

2.10 Teleology and Nature

If nature is spoken of in two ways—as matter and *ousia* (641a26)—then cause can be spoken of in terms of two categories (642a2): final cause or "the for the sake of which" and necessity (*anangkes*).[31] The concern of the earlier philosophers with matter gets translated into the necessary. The introduction of the necessary requires exploration. There are three kinds of necessity (two of which are said to be spoken of in the missing treatise *On Philosophy*). The third kind of necessity is restricted to things which are generated (*genesis*) (642a6). This kind of necessity has already been identified as hypothetical (639b25): if a living thing is to grow and remain alive, it is necessary that it have nourishment (*trophê*) (642a7). To illustrate this kind of hypothetical necessity, Aristotle uses a tool as an example: "Since an ax is made to cut wood, it is necessary that it be hard: if so, it is necessary that it be made of bronze or iron. Similarly, since a body is an organ (*organon*) (for each of its parts exists for certain purpose, and so does the whole body), it is necessary that it be of such a kind and composed of such parts, if it is to fulfill its purpose" (642a10).[32]

The faculties of the soul, according to the account in the *De Anima*, include, nutrition, perception and cognition. The lowest common denominator in all living beings is the nutritive soul: "first one ought to speak about food and begetting offspring, since the nutritive soul belongs also to the other living things and is the first and most common potency of soul, by which life is present" (*De Anima* 415a24–29). The capacity to take in food and incorporate it into the living body is what determines what we recognize as living. The nutritive soul, in its role of

---

[31]Gotthelf suggests, "to be 'teleological', all will agree, an explanation must have some such form as 'A is present/occurs because A is necessary or best from some end B.' Typically an 'end' is defined as a good outcome" (Gotthelf 1987, "Postscript" p. 231). For the view that the teleological account was given by Aristotle to draw attention to the material-efficient process (the "as-if" account) see Nussbaum (1978b).

[32]Balme recognizes the importance of the passage in which Aristotle indicates that the body as a whole, like any other instrument (*organon*), is for the sake of some complex activity (*praxis*) (*PA* 645b15–18) and stresses the role of activity in Aristotle's search for the causes (Balme, 1987b, p. 88). As Balme (1987b, p. 88) and Pierre Pellegrin (1986, p. 113) point out, the differentiae are distinguished within the *HA* under the headings of "parts, lives, activities, character" (*HA* 487a11). The activities of animals obviously make up their life history characters.

begetting offspring, could be said to point beyond the individual as well as towards the individual in consumption of food (e.g. *De Anima* 416b16–25).

With regard to nourishment and growth, Aristotle tells us that among the most complete creatures (*zoiois tois teleiois*) there are two parts that are most necessary (655b30)—that by which food enters, i.e. the mouth, and that by which residues exit, i.e. the anus.[33] In this formulation we have a mixture of the final (*teleos*) and the necessary: the final applies to the organism as a whole (*zoios tois teleiois*), while the necessary applies to the parts of that organism (*anagnkaiotata moria*) in the face of the need for nourishment and growth. If one were to take the mean between these two most necessary parts—mouth and anus—one would discover the principle of life (*hê archê estin hê tês zoes*) (*PA* 655b39). The doctrine of the mean applied to two holes yields the seat of life.[34] Though this may be meant as a joke, these two most necessary parts highlight the process of nourishment and growth, the struggle for existence and the maintenance of the living, organized whole. With the emphasis on the completed animal whole *(zoios tois teleiois),* including its nourishment and maintenance, we are able to view the phenomenon of life from a different perspective. What is discovered from this new perspective is that mouths are not simply instruments of nourishment, there is a plurality of functions associated with animal mouths so that there is a kind of freedom within this most necessary part.

Plants and animals take up food or nourishment (*trophê*), transforming it into material used in the service of maintaining the whole. Food, in the form of the solid (*zeron*) and fluid (*hugron*), is subjected to the hot (*to thermon*) in the process of concoction or change (650a5). The element fire is transformed into a principle of heat that does a certain work for the organism and accounts for certain aspects of generation.

The interpretation of Aristotle's treatment of nutrition and the feeding behaviors of different animal kinds is benefited by a close analysis of the biological particulars, including an examination of the different mouth morphologies that present themselves in different organisms including plant-like or generally bloodless animals like sea-squirts and cephalapods, fishes, birds and human beings. The discussion of the particular mouth morphologies and feeding behaviors complements the discussion of the nutritive soul, or at least the aspect of the nutritive soul devoted to food and feeding.

[33]Lennox (2001) suggests that his formulation "might refer to the mature stage of each animal, or it might imply a distinction between more and less 'complete' or 'perfect' kinds of animals. Aristotle is capable of making both points, and occasionally both at once, as at GA II.1, 733a33–b2)." In this investigation, I will argue for the former, but only if "mature stage" is not simply restricted to sexually mature individuals. For an alternative view that stresses the notion of complete or more perfect kinds, see Gotthelf (1987).

[34]Lennox (2001) observes that a similar observation is made elsewhere. Understanding the heart as the part that is a mean between mouth and anus helps understand why the issue of blood is of such importance in *PA* II.

The discussion of necessity originates, in part, with Aristotle's suggestion that the final cause and the beautiful are more fully at work in the works of nature compared to the works of art (639b19–21). Final cause and the beautiful are in a certain sense simply in the works of nature; necessity, Aristotle suggest in the context of the claim about final cause and the beautiful, is also present in the works of nature, but one must specify which form. In other words, there are distinctions to be made within what we understand to be necessary. The first sense in which we understand necessity, Aristotle tells us, is as simple necessity, which belongs to the eternal things (639b24). Aristotle's predecessors and others interested in such questions want to take their accounts back to necessity, they want to ground their own understanding in some view of the necessary. The necessity with which they are most concerned is probably the simple necessity, in so far as they want to say, for example, that "all is water" forever.

The second form of necessity is hypothetical or conditional necessity as understood in contrast to simple necessity. Hypothetical or conditional necessity is "present in all generated things, as it is in artifacts such as a house or any other such thing" (639b24–26). From the perspective of conditional necessity, matter has to be present in the beginning for any kind of generation to take place (639b26–30). So, for instance, the coming to be of the house requires an initial outlay of building materials. Included within the rubric of conditional necessity is the thought that the original matter is changed in the process of coming to be (639b28); in fact, Aristotle claims that matter is changed and changed continually until the end is achieved.[35] To give a simple example, one can see how an initial store of lumber must first be sized and cut before it can become part of a wall of a building. Again, conditional necessity points to the process by which matter changes and is shaped continuously as it approaches an end (639b26–30). The emphasis in this discussion is not on the eternality of matter, but rather on the flexible motion and changeable nature of matter when one recognizes coming to be or genesis. This form of necessity is in the beginning understood in light of the craft analogy; it's easy to see how the right tools and materials are necessary in the beginning. But one must move from the craft analogy to the application of conditional necessity in living things. How this plays out will be an important aspect of the argument of the *PA*, or in the argument of the *Generation of Animals*. Conditional necessity as applied to living things seems best able to describe the growth of an individual organism from a seed, which, as we'll see, is what this distinguishes it from the third kind of necessity.

The third form of necessity is set apart in the argument from a discussion of the first two sorts. The third form of necessity is, "as it were," practically conditional necessity (642a8). Like conditional necessity, the emphasis is put on genesis and those things that partake of generation. But the discussion of this third form of necessity seems to focus on nutrition and nourishment (642a7). The continual maintenance of the whole animal or complete living being is the phenomenon by

[35]Cooper (1987) stresses that, for Aristotle, hypothetical or conditional necessity is always understood relative to an end (pp. 243–244).

which this third form of necessity is most applicable.[36] Something like metabolism seems best able to capture what Aristotle recognizes.[37] The short discussion of this third form of necessity culminates in Aristotle's suggestion that the body is an instrument (642a11); the job of an ax is to split wood, so it must be composed of something hard and, one might add, be a certain shape. The work of the organic whole, whatever it might be, requires a proper instrument, a metabolic instrument one might say. Again, this third form of necessity seems to imply the process of the maintenance of the whole, but also can expect a discussion of Aristotle's response to Democritus that the parts, and I would add the whole organism, serve some function or do some work (640b29–641a4). Conditional necessity is best exemplified by ontogeny, whereas the third form of necessity is best exemplified by the process by which an animal assimilates nourishment. The difference between conditional necessity and this third form, what one might term "metabolic necessity," mirrors the split in the nutritive soul (*De Anima* 415a24–29).

A certain understanding of necessity turns living bodies into tools.[38] The need for nourishment by living things provides the first example of how we are to think about this form of necessity (642a7) in that nourishment is necessary as a condition for the body or organism to fulfill its purpose. If an organism's purpose is in its living, then its way of life is that which directs the organization of the *trophê* into parts. The need for nourishment dictates the necessity of certain parts, in a certain arrangement. The "for the sake of which" necessitates the structure of the body; that is, the struggle to live organizes matter. Matter alone does not organize itself, as is implied by Aristotle's predecessors (640b13). Air or water cannot be both material cause and efficient cause. Even if they could, the fact would not explain the origin of life.[39] Consideration of the relation between teleology and necessity propels one to at least three problems in biology: how the animal form is shaped; how one accounts for the origin of life; and how form is reproduced by an individual in its offspring.

---

[36] I understand the "two modes defined in our philosophical treatises" (642a5) to be simple and hypothetical necessity as discussed earlier in *PA* I. For alternative views, see Balme (1987c, p. 285) and Cooper (1987, pp. 259–260).

[37] Freudenthal (1995, p. 182) seems to recognize something like this in his treatment of what he terms the "physiological theory of the functions of nutritive soul."

[38] See also the textually difficult description at *PA* 662a23 ff. where Aristotle talks of the differentiated workings (*tês egasias diaphoras*) of the organism. Additionally, in a prelude to the discussion of the variations that occur in uniform and non-uniform parts, Aristotle tells us that the uniform parts contribute either to the *ousia* of the non-uniform or to the function of the instrumental part (647b25) and that the differentiae are explained as being required for the functions (*ta erga*) and the *ousiai* of the animal (648a16; cf. Gotthelf 1987). In these instances the differentiae and *erga* are juxtaposed in a way that might prompt the question, are the differentiae meant to be seen as *erga*? Additional evidence for the view I am arguing for is indicated by Aristotle's discussion of instances when *ergon* seems to take precedence over the necessary cause in the formation and organization of the material of certain parts (658b23–26).

[39] There has to be a point at which matter pre-dates life. The assumption of self-organizing or living matter is Buffonian.

In the *Physics*, Aristotle shows why the dynamic between teleology and necessity can best be seen in the arts:

> Is the necessity 'on a hypothesis,' or is it in fact absolute? Some suppose that the necessary exists in things that come to be in the way that one might think that a wall had come about necessarily because heavy things naturally go down and light things go up, so that the stones and foundations went to the bottom, the soil above them, and the timber to the top as being the lightest. Now the wall did not come about without these things, but nevertheless it was not because of them, except as matter, but for the sake of giving protection. Similarly with everything else that has a final cause: it is not without things that have a necessary nature, on the other hand it is not because of them except as matter: the necessary is in the matter, but the final cause is in the definition... clearly, then, the necessary in natural things is that which is spoken of as matter and movements in it (*Physics* 199b34).

In the construction of a wall, it is necessary that the heavy things go on the bottom, but it is not necessary, in the same way, that the wall be constructed at all, except as it relates to the purpose of defense or protection. This is an example of the interaction between purpose and necessity.[40] Some purpose (wall as protection) interacts with necessity (some characteristic of the elements that requires heavy things to be placed on bottom) in formation.[41] In one sense, the cause of the wall's generation is a result of the goal of protection. The wall will be built in such a way as to protect, e.g. not a two-foot high structure built of paper mache. But one can also point to necessity as a cause of the wall's having the character it does; the construction of the wall to serve for protection is limited by the characteristics of the material elements involved in construction. At the organismal level, if the goal or purpose is living, the form of the body will develop in such a way as to serve the purpose, while being limited by the characteristics of the materials that form the animal body.[42] In addition, these materials are necessary if that form is to be

---

[40]While we do not have to agree with the statements regarding the difference between Plato and Aristotle, the relationship between the useful and the search for causes is nicely expressed by Balme (1987c, p. 277): "Aristotle too expresses the distinction between the causes as between necessity and 'the better' or the 'good', although he makes it clearer than Plato does that 'good' is not an extrinsic value-judgment but means the useful or advantageous from the animal's viewpoint." Compare also the passage at 639b19 where Aristotle suggests that final cause is more present *in* the works of art than in the works of nature.

[41]Compare Balme's suggestion: "The directiveness that Aristotle sees in nature is more than natural interactions, so that the teleological explanation coexists with the causal explanation. But he bases the teleology not primarily on directiveness but on the existence of forms. To explain an organ, he says, you must first grasp the complete animal's form and functions, what it means to be that animal, its *ousia*" (*Dictionary of Scientific Biography*, p. 259). Compare this to the view that *ousia* is somehow moving and formal cause (641a26). Balme does recognize the importance of function in the examination of *ousia*.

[42]Gotthelf (1987) protests that Aristotle never really explains how something comes to be for an end. He suggests that "readers of the corpus will search in vain for a detailed analysis of what it is to be (or come to be) for the sake of something" (p. 204). To examine teleology, not in the case of the generation and development of an animal, but in the case of a functioning being in its environment (cf. Gotthelf 1987, esp. pp. 207–208), one must examine the habits and activities of animal wholes.

possible. Yet, as already noted, necessity can have a number of different meanings not limited to material causality.

While it is not clear that he understands his agreement with Aristotle on this account, D'Arcy Thompson gives some evidence to support the *PA* position:

> [L]ike warp and woof, mechanism and teleology are interwoven together, and we must not cleave to the one nor despise the other; for their union is rooted in the very nature of totality. We may grow shy or weary of looking to a final cause for an explanation of our phenomena; but after we have accounted for these on the plainest principles of mechanical causation it may be useful and appropriate to see how the final cause would tally with the other, and lead towards the same conclusion (*Growth and Form*, p. 7).

The *PA* will weave together in its account of the parts of animals the two perspectives of the necessary and teleological. The nature of these things cannot be understood without these two explanations together. To present the teleological and the mechanical definitions is to combine the ways of the dialectician and natural scientist: the dialectician may describe a feeling or *pathos* like anger as a certain desire for retaliation; anger can also be described in mechanical terms, by the rise in temperature and boiling of the blood around the heart (*De Anima* 403a26). One goal of the *PA* as a whole is to examine the complementary as well as the apparently contradictory way in which mechanistic and teleological descriptions unfold.

Aristotle ends *PA* I.1 with a discussion of respiration that is, at first, hard to understand in light of what has come before (642a32–642b4). This discussion of respiration highlights the explanatory power of both the necessary as a cause and the end. The argument progresses from the mention "that breathing exists for the sake of this"—where "this" is understood as the functioning of the animal whole—to a discussion of the matter and mechanics of breathing. We are to incorporate the necessity bound up with matter, the elements and their motions with an eye towards animal wholes. In this regard, Aristotle claims that "it is necessary for the hot to go out and enter again upon meeting resistance, and for the air to flow in. This is directly necessary; and it is as the internal heat retreats during the cooling of the external air that inhalation and exhalation occur" (642a35–642b3). One could say that the cause of animal respiration is explained by the lungs functioning for the sake of the whole. But as is suggested in this final passage, one could also say that the cause of animal respiration is to be found in the motion of hot and cold air. This is to begin to weave together the mechanical and the teleological.

Aristotle's pre-Socratic predecessors provided a way to think about elements while the "Socratics," in addition to turning toward practical virtue and politics, inspired interest in the "what it is to be" (*to ti ên einai*) and substantial being (*ousia*). In the discussion of respiration that ends I.1, Aristotle hints at the way in which he will confront this divide head on. This discussion of respiration and the interplay between the necessary motions of the elements and final cause, anticipates, in addition to many other things, the better known example of the shark's mouth. It also anticipates the way in which *PA* II is structured, as we shall see. So, this brief discussion that ends *PA* I.1 sketches, in rough outline, themes that will reemerge throughout *PA* II–IV. The seeds for the whole book are sown in *PA* I.1.

Chapter 3
Recognizing Sameness and Otherness in Animals (*PA* I.2–5)

Abstract As Aristotle lays the grounds for his inquiry into the parts and lives of animals, he turns to a discussion of dichotomous division. As we shall see, the relationship between the examination of parts of a whole and the attempt to determine kinds either through dichotomous division or some other way proves difficult, but also provides an opportunity to broaden the discussion. We must attempt to see how what appears to be a digression on a method practiced in the Academy relates to questions and issues raised in *PA* I.1, in addition to illuminating certain issues one encounters later in the *PA* in the examination of the parts and lives of animals.

In the discussion of dichotomous division, Aristotle may have in mind the problem of being and nonbeing which runs through the discussion of Plato's *Sophist* and *Statesman*. How things are essentially different is connected with the problem of non-being. The discussion of dichotomous division expands, at the same time, on the topic of sameness and otherness, which we confronted in the earliest concerns about whether to proceed by examining how the same attributes occur in different kinds, in things which are other (e.g. 639a19, 639a27). If we recognize how that examination expands on the problem of the same and other—in addition to the relationship between the common and particular—we will be in a better position to appreciate the movement of *PA* I and that of the book as a whole. In reading *PA* I.2–3, we should bear in mind that Aristotle is reforming the way in which we view the same and other—which culminates in his discussion of the more and the less and analogy (644a14 ff.). By doing so, we make the apparently pedantic discussion of dichotomous division more illuminating for the larger argument. The wings of birds and the pectoral fins of fishes are used to illustrate elements of the philosophical argument regarding the problem of same and other.

The *PA* begins by dividing sciences (639a ff.)—a discussion that seems to proceed dichotomously. When the argument turns to the division of animals it also makes a definite turn to dichotomous division. When looking at Chaps. 2, 3, and 4 of Book I,

J.A. Tipton, *Philosophical Biology in Aristotle's Parts of Animals*, Studies in History and Philosophy of Science 26, DOI 10.1007/978-3-319-01421-0_3,

someone who assumes that there is some unity to the argument of *PA* I and the book as a whole must ask how the discussion of dichotomous division relates to the examination of the parts of animals. At a simple level, we differentiate kinds on the basis of parts; for example, birds have wings as parts which distinguish them from other organisms. Thus, we learn that there are some parts which might be part of a thing's substantial being in the course of the discussion of division into two (643a4). However, the relationship between the examination of parts of a whole and the attempt to determine kinds either through dichotomous division or some other way proves a bit more complex. We must attempt to see how what appears to be a digression on a method practiced in the Academy relates to questions and issues raised in *PA* I.1, in addition to illuminating certain issues one encounters later in the *PA* in the examination of the parts and lives of animals.

In addition to exploring the issues of sameness and otherness from the perspective of dichotomous division, one surprising way in which the problems arising in *PA* I.2–4 are to be connected to the larger discussion is through a reflection on speech and writing. This concern with speech is to explore aspects of human cognition at the same time one is examining the parts and lives of animals. If we think of the method of dichotomous division as an attempt at a kind of taxonomy—which, as we shall see, is an inadequate understanding of the argument in certain very important ways—Stephen J. Gould offers a description which can be fruitfully applied to our study of the *PA* in illuminating several areas: "When we recognize all influential classifications as careful descriptions of organisms made in the light of the fruitful theories about the causes of order, we can finally appreciate the fascination of taxonomy as a source of insight about *both* mind and nature" (*Natural History*, September 2000, p. 22). The *PA*'s discussion of dichotomous division is a reflection on the way in which humans view, and then attempt to classify, the organic world.

It is often taken for granted that the *PA* discussion of dichotomous division refers to passages in Plato's *Sophist* and *Statesman*.[1] In these dialogues, the relation between part and kind seems to indicate an important problem (*Sophist* 254c, *Statesman* 263b). In the opening of the *PA*, it appears as if Aristotle presents part (*meros*) and form (*eidos*) as synonymous: "It is evident that, if we consider each part (*meros*) separately, we shall often be repeating the same discussion" (*phaneron d' hoti kai kata meros men legontes peri pollôn eroumen pollakis tauta*) (639a 23).[2]

---

[1]Balme (1987c) suggests that both Plato and Aristotle were not interested in classification per se, but in getting at the definition of a thing, to "discover what exactly it is" (*heurein hoti pot' estin*) (*Sophist* 221c). One gets to this point, Balme suggests (p. 73) that as "the naming of genus with differentia will denote a single thing, the unified substantial *tode ti* which for Aristotle is the object of definition."

[2]While I generally appreciate Balme's translation of *PA* I, I believe he ignores a real problem when he translates this line as follows: "If we do speak of the animals severally, it is plain that we shall often be saying the same things about many of them." Thus, Balme (1987c, p. 71) is able to say that Aristotle reformed the Platonic view, which did not maintain a firm distinction between *genos*, *eidos* and part (*meros*). Evidence for such a "reform" is not provided, to say the least, by the passage at hand.

Most translators use "species" here for *meros*. The suggestion seems to be that part stands to whole as *eidos* (form or kind) does to *genos*.[3] The idea that animal kinds or their attributes are strictly speaking, parts of the larger whole of the animal genus, like a hand is part of the body, has important implications for our study of nature including how we understand teleology in nature and the related question of cosmology. If forms of animals or species are parts of a genus as a larger whole, those parts would presumably function for the sake of that whole. Does Aristotle view species or kinds, then, as mere parts? But what, then, of the individuals that make up those kinds? Do the individuals of different species have a certain wholeness of their own?[4] Perhaps, on the other hand, it is attributes, things like growth, decay and movement, that should be viewed as parts of a whole, a whole that describes organic life. I would suggest that the attributes (*ta sumbebêkota*) that came up in the earliest problem confronted (639a18), become the differentia (*ta diaphora*) in the discussion of dichotomous division. But as we shall see, there are no common attributes allowed in the practice of dichotomist division that is the subject of Aristotle's criticism.

In addition to the problem of the relationship between part and kind, Aristotle may have in mind the problem of being and nonbeing which runs through the discussion of the *Sophist* and *Statesman*. Aristotle indicates this associated problem in his discussion of differentiation and the need for incorporating privation (642b11). How things are essentially different is connected with the problem of non-being. The discussion of dichotomous division expands, at the same time, on the topic of sameness and otherness, which we confronted in the earliest concerns about whether to proceed by examining how the same attributes occur in different kinds, in things which are other (e.g. 639a19, 639a27). Between posing the question of whether to proceed by taking each being separately or by examining common attributes (639a16 ff.) and the apparent answer to this question (e.g. 645b1, 645b19), we examine dichotomous division. If we recognize how that examination expands on the problem of the same and other—in addition to the relationship between the common and particular—we will be in a better position to appreciate the movement of *PA* I and that of the book as a whole. In reading *PA* I.2–3, we should bear in mind that Aristotle is reforming the way in which we view the same and other—which culminates in his discussion of the more and the less and analogy (644a14 ff.).

---

[3]In this context, it is worth noting one way in which Pellegrin (1986) understands the terms that might allow one to see the emergence of the modern taxonomic project: "Aristotle thus conveys by the term *genos* the transmissible type that in our eyes characterizes the species, and by *eidos* the model that is actually transmitted in generation. It would be necessary for these two notions to converge and become superimposed for the modern concept of a species to be born. For Aristotle, the species did not yet exist" (p. 110). This is a continuation of Pellegrin's argument to the effect that neither of the terms, *genos* or *eidos*, "indicates a constant degree of generality on which a taxonomic construction could be based" (p. 106).

[4]The issue of the way in which species are defined and its connection to the question of the relation between part and whole is still very much a problem for contemporary systematics. This is evident in Michael Ghiselin's (1974) radical solution to the species problem, which suggests that species are composite wholes, that they are individuals in an ontological sense.

By doing so, we make the apparently pedantic discussion of dichotomous division more illuminating for the larger argument.

Our inquiry into nature is directed at discovering the whatness of natural things, and in particular, the whatness of living things. Thematic issues in *PA* I.2–4 include privation and a discussion of what it means to be same or other. These issues are embedded in an analysis of the method of division. Dichotomous division provides one model for discovering "the what is" of living things. In differentiating or distinguishing one living thing from another, one would expect it to attempt to get at a thing's being.

3.1 Dichotomous Division

The dichotomous method involves taking a one, a *genos*, and dividing it into two differentiae (642b5), just as the whole of number is divided into even and odd or the whole of humanity into male and female. However, according to Aristotle, limiting the division of a *genos* to two is difficult, if not impossible (642b6). Of course dichotomy divides a *genos* into a many by means of a succession of pairs of terms; thus, in the end we collect the vertical elements of our pairs, such as A and B and D or A and C and F in Fig. 3.1. We will have to see if the practitioners of this method, as Aristotle presents them, allow such a procedure. Aristotle wants to ensure that differentiation proceeds by a means other than division into two. But this concern with rejecting dichotomy seems also to be a concern with preserving opinion, or at least beginning the inquiry from the point at which groups are delineated in everyday language.[5]

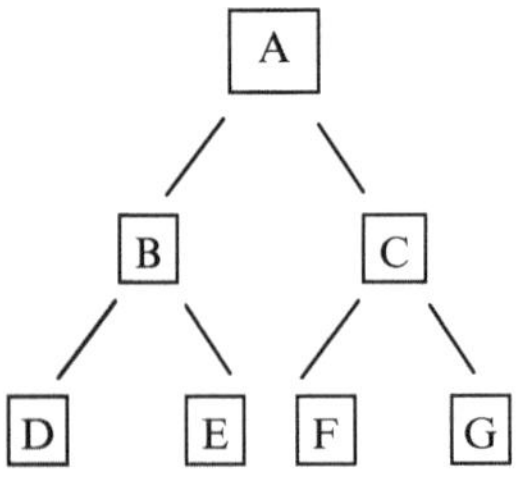

Fig. 3.1 A representation of dichotomous division

[5]Pellegrin (1986) discusses the problems commentators see with Aristotle's apparent inability to break with contemporary language. This discussion is part of Pellegrin's larger argument against a tendency in commentaries concerning Aristotle's classification of animals. He notes: "Commentators notice Aristotle's obvious and explicit desire to separate animals into nonarbitrary groups, and they add a presupposition, so indisputable in their eyes that they do not formulate it: namely, that Aristotle had to be trying to achieve that 'perfect' classification, the binominal taxonomy, which we call, perhaps erroneously, Linnaean" (p. 6). Pellegrin identifies that presupposition and argues against its insertion into Aristotle's thought.

Aristotle seems to offer his own version of *diaphorein* as an alternative to the method of division that proceeds dichotomously. The method of division is one half of dialectics, understood as collection and division. If *diaphorein* is meant to replace dichotomous division, we must wonder whether Aristotle also has a corresponding replacement for collection. Otherwise, his own version of differentiation would entail both collecting and dividing, making *diaphorein* the equivalent of dialectics applied to living organisms. If differentiation is to be understood as akin to dialectics, comprising both collecting and dividing, differentiation obviously should begin with opinion, as Aristotle indicates.

Aristotle levels at least six criticisms against the method of dichotomous division, but there are some features that should be highlighted.[6] The first criticism leveled against the practitioners of dichotomous division seems to be motivated by a concern for speaking: "For of some things there will be only one difference, the others being superfluous, e.g. footed, two-footed, split-footed; this single difference is decisive. Otherwise, it will be necessary to say the same thing many times." (642b7–9). The "digression" on dichotomous division, as this criticism shows, is in fact connected with Aristotle's concern that certain common attributes will be repeated many times (639a25). The problem of saying the same thing over and over reflects the fact that some attributes occur in many, if not all, living things.

The example that Aristotle chooses to illustrate this criticism is not easy or straightforward. The characteristic of being cloven footed, a foot split into two toes, is not restricted to those animals that are two-footed. This means that the vertical elements in the dichotomy have to be combined in order to classify a given organism, for example, a two-footed, cloven-footed organism versus a four-footed cloven-footed organism. If, however, the assumptions and the aim of those practicing dichotomous division do not allow for the collection of all the elements employed in the division, the "method" is purely analytic with no synthesis involved. Aristotle's turning to the act of collecting that is behind the everyday names applied to groups of organisms is an attempt, in that case, to highlight the absence of synthesis within dichotomous division. It also might be hypothesized that the term "*eidos*" points to the act of division while "*genos*" illustrates collection; evidence for this will have to be found within the discussion of dichotomous division and the process behind popular designations.

If the vertical elements of any given line of dichotomous division have to be included in the classification of an organism, then those higher elements are not superfluous, as Aristotle claims. Two-footed is not at the bottom of the dichotomous line, but must be included in the description. Aristotle's criticism seems to be that taking all the vertical elements in a dichotomous division is too lengthy a process, requiring things to be said many times. This issue returns in *PA* I.3 when Aristotle suggests that the manner of expression in dividing dichotomously "makes it seem" as if the final differentia is the only one (643b35). The deficiency concerns the manner of expression of dichotomous division; as we shall see again, the practitioners of such a method do not understand speech (Figs. 3.2, 3.3, and 3.4).

---

[6]Balme (1987b) sees eight criticisms or rules for division.

Fig. 3.2 Aristotle discusses bird wings and fish fins at length in the early chapters of the *PA* highlighting the ways in which they are analogous. Elsewhere, he distinguishes three types of wing—the feathered, the membranous and the dermatous (*HA* 490a6). The dermatous wing is the wing that characterizes the bat

Fig. 3.3 The membranous wing of a dragonfly is characteristic of the bloodless, in contrast to the feathered and membranous wings of those animals with blood (*HA* 490a9). The above is an example of a four-winged, bloodless, unsheathed insect that, Aristotle says, is relatively large (*HA* 490a18)

An additional criticism Aristotle levels against the method of dichotomous division also seems to involve speaking. It is a mistake, Aristotle tells us, to break up a group, a *genos* like birds (642b11), placing some birds in the water creatures and others in a different class. When trying to determine a certain kind

Fig. 3.4 The wing-like pectoral fin of a gurnard (*Dactylopterus* sp.) in a fish market in Mytiline (*top*). In his *Progression of Animals*, Aristotle explains that "birds in a way resemble fishes. For birds have their wings in the upper part of their bodies, fishes have two fins in their fore-part; birds have feet on their under-part, most fishes have fins in their under-part and near their front fins; also, birds have a tail, fishes a tail-fin" (714b1–b6). One can note the relative placement of the pectoral fin (*top arrow*) and the pelvic fin (*bottom arrow*) in the goby (*bottom*). It is this relative placement that Aristotle is emphasizing in the comparison with bird wing and legs

of hunting, Plato's Eleatic Stranger suggested a certain kinship between those organisms that are feathered and those that live their lives in water (*Sophist* 220a7)—he groups fishes and birds. Such a grouping is made on the basis of a recognition of the similarity of movement between some types of fish and some types of birds (Figs. 3.5, 3.6, and 3.7). Grouping fishes and birds puts emphasis on the similarity of

Fig. 3.5 A flying gurnard spreading its pectoral fins, it's "wings," so to speak. Fish like this and the rays and skates are examples that allow one to see how birds and fishes could be grouped together. Gurnards are known to make short gliding leaps out the water (Luther and Fielder 1976). In talking about the sound that creatures like this make in flight, Aristotle says of the flying fish (probably *Exocoetus volitans* according to Thompson (1947, p. 286) that "this fish flies quite clear of the water, without touching it, having long broad fins" (*HA* 535b27)

Fig. 3.6 Great blue heron gliding over the water

Fig. 3.7 A sting ray gliding through the water. Again, one can see the resemblance with birds with respect to their motion. Aristotle says of the rays that they "swim by means of their broad bodies merely" (*HA* 489b30). He says also that fishes like the rays and skates have long and spinous tails (*PA* 695b10)

parts—in this case wings and pectoral fins (Fig. 3.4)—and the way in which they are used in motion—swimming and flying. While this seems to be an instance of seeing the same or similar attributes in the other, in different kinds of animals (639a28), it abstracts from obvious differences, the otherness of fish and birds presented to the eyes. Such a grouping also abstracts from the element each inhabits, the aqueous or aerial. Despite English expressions like "fly through the water," birds and fish have group names attached that identify them as distinct (642b14); the distinction between bird and fish as groups is determined popularly (643b12). The popular grouping of organisms, Aristotle suggests, has to be the starting point. Perhaps the starting point for this kind of inquiry can also be viewed in terms of the groupings given with everyday use—the inquiry starts with everyday opinions about living organisms. The group "birds" is held to be a kind of whole, from which one can further discriminate parts or kinds of the whole.

To group fish and birds based on the similarity of movement is to go against the general opinion of such things. It is also to go against what is given to us by sight. Birds and fish obviously look different, having different morphologies. However, morphology can deceive. Morphologically speaking, a dolphin is much more similar to a shark than almost anything else. But unlike sharks, it is difficult to categorize dolphins simply as "aquatic"; recognition of this causes our scheme to be significantly revised (*HA* 589a13). If the conventional differentiation of birds and fish were based on form or shape alone, the practice would be open to the same criticisms that Aristotle levels against Democritus' reliance on shape alone (640b25 ff.), or at least the Democritus that Aristotle presents to us. One can begin with form or shape in beginning with opinion, but it may prove necessary to go further.

The designation of birds and fish also may be the result of a crude sort of elemental analysis. The medium of birds and the medium of fishes should not mix, thus the organisms inhabiting such elements should not mix. Aristotle's criticism of the mixing of fish and birds is complicated further, by his own description of the flying of certain fishes (*HA* 535b27). If Aristotle begins with the groupings of animals based on opinion, it is probably safe to say that, depending on his purpose, such groupings may be modified. In grouping fish and birds, the dichotomists neglect opinion as a starting point, but perhaps more importantly they seem unconcerned with synthesis, which is required for an examination of the causes of the parts and lives of animals.

3.2 Nonbeing

In light of the themes that dominate Plato's *Sophist* and the *Statesman*, it is not surprising that Aristotle's discussion of dichotomous differentiation introduces the problem of nonbeing. In trying to determine how one kind is differentiated from another kind, one has to resort to employing terms like "same" and "other." Otherness relies on a thing *not being* the same as another thing. In this sense, the notion of privation is meant to express non-being. However, based on the assumptions and principles of the method of dichotomous division according to Aristotle, privation should not be employed in dichotomous divisions. This is due to the fact that there are no *eidê* of nonbeing (642b23); for instance, there cannot be forms of footlessness.[7] Generic differentiae or differentiae on the whole (*katholou diaphoras*) must contain *eidê* in order to be generic or apply to the whole (642b24). Privation can be employed by the dichotomists, but only at the end of the line of division; nonbeing can be used to distinguish the particulars at the last stage of the divisions, according to Aristotle's description of the practice. Of course, this results in the paradoxical predicament of privation being the final and definitive "character" of a given kind.

[7] While Balme (1987b, p. 75) gives an example of dividing footless into snakes and fishes, he insists that "what we are dividing is not footless but those animals that are footless." This illustrates the move to separate the animals and their attributes that will culminate in *PA* I.5 (see discussion below). This suggests to me that the dichotomists do not go in the direction of Aristotle in seeing the need to divide the attributes from the animals that are being studied. Thus, they cannot divide by privation since that would highlight the distinction between character and animal, e.g. there are animals that are footless, but according to the dichotomists there cannot be *eidê* of non-being within their classificatory scheme.

There are some peculiarities in the examples Aristotle provides to support the argument that privation cannot yield division into species. We can see how featheredness could be divided further into unsplit (*aschiston*) and split (642b29), while featherless is a terminal node. In the case of footedness, we can divide along lines of cloven and uncloven (*aschides*) or undivided (*adiairetos*). Cloven can be divided into many parts, as in the feet of humans, or into two parts, like those of sheep. Animals with undivided hoofs include things like horses and mules, but also perhaps certain pigs (*HA* 499b11), which seems to illustrate a case in which a form of privation—the undivided (*adiaretos*) hoof—divides into species.

Many of the examples of privation seem to highlight some aspect of the not-divided kind. The method of division yields the undivided element or character, not-split feather, not-divided hoof. Divisions are supposed to terminate in that which cannot be divided further, presumably the *eidos*. In the case of Aristotle's examples, dividing eventually runs into something that cannot be further divided, which turns out to be the privation or nonbeing of the split. An undivided (*adiairetos*) hoof is, at least in some sense, the nonbeing of the split hoof. In other words, privation seems to indicate things that are not divided, things which are, in a sense, whole. The divided or the split turns out to be the primary phenomenon, with the unsplit or undivided being derivative of it.

Aristotle introduces privation in the discussion of dividing animals into kinds, but now he introduces a new element in the investigation by applying it to animals themselves. Ants can be winged or non-winged, depending on their stage of life (642b34). Perhaps Aristotle's point is that some members of the same species might have wings, while others have none. If an animal like the ant is winged and wingless at different times of its life, it will be included in more than one line of dichotomous division, one that includes winged things and the one that does not. Dichotomous division, in that case, could not account for things which change over the course of time, from one way of being to another.

How does talk of privation, as a form of nonbeing, relate to an examination of the parts of animals? If kinds are to be understood as parts of a greater whole, this helps make some sense of the appropriateness of nonbeing in this discussion. But how could nonbeing relate to the organism? It would be very strange to talk of nonbeing as a part of an animal. An animal is defined by the parts it has, not by what it does not have. How can something be characterized by what it does not have? Unless what the thing does not have is divisions within itself; perhaps paradoxically, privation can offer a way to talk about something's wholeness as not having splits or divisions. This possibility anticipates the move in the argument to a discussion of indivisibility (*atomas*). Of course, a whole of parts may not be the same as a whole that is indivisible. In any case, there is the problem of what we take the whole to be in the context. We must decide whether to focus on animal life as a whole, or on the kinds (*eidê*), or the individual organisms.

3.3 The Indivisible Species of Being *(eidos ti tês ousias atomon)*

The differentiae, as applied to the method of dichotomous division and, as we shall see, Aristotle's own method, mark off kinds. However, in Aristotle's formulation, the differentiae of dichotomous division are inextricably tied to particular lines, leading to an identification of differentia and *eidos*. The equating of differentia with *eidos* is suggested by Aristotle's sliding from one to the other (643a1–8). Aristotle indicates that as the dichotomist proceeds, particular differentiae will define particular kinds. They will thus be unavailable in different parts of the tree (for example, the character that defines "G" in Fig. 3.1 cannot be used to move through the line that terminates in "D"). Behind this criticism of the dichotomists is a desire to break the association between differentiae and *eidê*. Aristotle hopes to be able to utilize the same differentia, in combination with others, to identify disparate kinds. This desire to break the isomorphic relationship between *eidos* and differentia results, in part, from a recognition of the attributes that remarkably different animals can share (e.g. 639a27–b6). Aristotle thus indicates that those committed to the assumptions of dichotomous division are not in a position to recognize and appreciate the significance of common attributes. Hence, they do not see how the common (the same) becomes the particular (other) in different kinds of animals.

The model of dichotomous division presents the last differentia as an indivisible (*atomos*) species of being (643a2). At this point the "indivisible species of being" appears to refer to the descriptive differentia. This situation, however, changes or at least becomes unclear. Aristotle stresses two related points. First, he is concerned with being able to use different differentiae in delineating different kinds, which requires denying the linkage of differentiae and *eidê*. One differentia belonging to more than one kind would appear to be a violation of dichotomous division (643a6), since the differentiae are indivisible, which seems to mean not in common. So, for example, one would like to be able to use the character two-footed in many different ways. In the dichotomist's framework this would not be allowed if the descriptive were tied to one particular kind.

Aristotle's second point targets the dichotomist's desire to reduce one *eidos* to one differentia. The dichotomists want to determine an atomic unit, which corresponds to one particular descriptive differentiation. This may be more of an artifact of the method than a reflection of the nature of things. A kind, such as man, must necessarily have many differentiae that marks it off (644a7). But is this to substitute the method of dichotomous division with opinion or popular distinctions (643b13)? This would be an unsatisfying substitution, but it could be Aristotle using opinion, again, to correct elements of the dichotomist's way. In the course of the argument we have moved from privatives like not-split feather (642b29) to the unsplitable atomic units of being (643a1).

That there is a one-to-one correspondence between differentiae and indivisible animals (where "animals" is used synonymously with *eidê*) is demonstrated in the dichotomist's way of proceeding by the fact that there will be an equal number of differentiae and kinds (643a7). The dichotomists aim at a numerical correspondence between differentiae and kinds in their determination of animal taxa. In order to preserve such a numerical correspondence the differentiae must be peculiar or particular and not common (643a12). From the perspective of dichotomous division, common differentiae lead to a confusion of same and other; with the identification of differentia and *eidos*, the dichotomist falls prey to the principle of non-contradiction if he uses common differentiae. This again seems to point to an inadequate understanding of the same and the other on the part of those practicing dichotomous division; making the last differentiae peculiar is a way in which to ensure that the other does not enter into the same (643a13).

Aristotle objects to the identification of one differentia and one *eidos*. As an alternative, he puts the analysis in terms of parts: "It is the differentia in the matter that is the *eidos* (*esti d' ê diaphora en tê hulê to eidos*). For just as there is no part of an animal without matter, so there is none that is only matter; for it is not body irrespective of state that can be an animal or any part of one, as we have often said" (643a24). Just as an animal is composed of a plurality of parts, there will be many differentiae. The notion that a group is defined by many differentiae is first rooted in opinion or the popular designations (643b12). Yet differentiation, which is ingredient in a division of kinds, is connected with an analysis other than popular designation, namely, an analysis in terms of parts and wholes.

Aristotle has been suggesting that the dichotomist's designation of *eidos* by one differentia is not adequate. He thus pushes us towards the use of a plurality of descriptive differentiae. This direction is also indicated by the seeming identification of differentiae and parts; on such a model, many differentiae should make up a designation, just as many parts make up a whole. Dichotomous division yields one differentia that corresponds to the *eidos* and Aristotle's criticism of the method leads us to believe that this drive toward a one is a failing. However, such a criticism does not seem consistent with Aristotle's assertion that it is **the** differentia that is **the** *eidos* (643a24). Could Aristotle, like the dichotomist, be looking for one differentia in trying to determine kind? What would that one differentia be? A possible hint is given when Aristotle connects what he says with regard to differentiae and *eidê* (643a 27) to a discussion of soul (641a18 ff.). We were under the impression that we were looking for a plurality of differentiae, but at the point Aristotle says that the *eidos* is the differentia in the matter we are left with the suspicion that we are looking for soul or something very similar to soul.

The discussion of soul (641a18 ff.) as differentia makes one wonder whether dichotomous division is in some way unable to account for soul, as it appears it is unable to account for nonbeing as privation. In suggesting that it is not "body irrespective of state" (643a26) that determines the animal and its parts, Aristotle suggests that one of the big deficiencies of dichotomous division is its over-reliance on appearance and matter. This would make Aristotle's task very difficult since he uses the *genê* marked off by poplar usage, which rely mainly on shape of the body (644b7), to correct dichotomous division, which relies on matter and shape of the body. Things can look the same and be quite different or can look very different but have much in common. As we shall see, both dichotomous division and the popular designations fail to capture this complexity.

3.4 Dividing by Ousia

Aristotle moves from a criticism of the identification of differentia and *eidos* to a concern that we divide by what is in the *ousia* and not by accidental properties.[8] The example he gives is taken from geometry: one would fail to divide in accordance with *ousia* if "one were to divide figures (*schemata*) on the basis that some have their angles adding up to two right angles and others to more; for it is an accident of the triangle that it has its angles adding up to two right angles" (643a 29). We are told to divide by things in the *ousia*, but we must ask whether *ousia* refers in this analogy to "figure" (*schemata*) or to triangles? If we were to divide the genus of figure into those which have angles adding up to two right angles and those that do not, we would not be dividing down the middle of the genus. Such a division would take a tiny part, namely, triangle, and set it against the much larger class of the remaining figures.[9] If things are to be divided down the middle, it is usually the case that opposites will yield such division. For example, perhaps one could divide a group by the straight members and the curved members (643a32). This will also ensure that divisions are made by similar terms avoiding situations in which different groups are divided by different characteristics—for instance, dividing one group of animals on the basis of swimming and another on the basis of color (643a34).

3.5 Common Functions of Body and Soul

Dichotomous division can be employed in trying to determine the whatness of many things, including geometrical figures, the arts, and number. But with regard to living things, which are ensouled things (*empsuchê*), dichotomous division seems to be unfit to account for functions common (*tois koinois ergois*) to body and

[8]As Lennox (2001) notes, this presupposes that one possess knowledge of the *ousia* in advance (p. 162).

[9]I take it that this is analogous to the division of the human genus into Hellenic and barbarian (*Statesman* 262d, 263c). In addition, if one were to divide the whole of number into 10,000 and the rest, one would not be dividing correctly. In these cases, a small part would be playing an inordinately large role. The Eleatic stranger in the *Statesman* suggests that it would be more beautiful to divide down the middle, to divide by two, so that we split number into even and odd, and humans into male and female.

soul (643a36).[10] The concern that Aristotle has for the common, whether the common attributes or the common functions of body and soul, keeps emerging in the discussion of dichotomous division. Like the Democritean reliance on form that Aristotle presents, dichotomous division works best by disregarding whether a thing is alive or dead. In addition, the mathematical character of dichotomous division would seem to make it unable to deal with bodies in motion, with living things. Animal locomotion requires both body and soul: the soul is understood as the principle of motion. Locomotion reflects an organism's psychic world coupled with its way of movement based on its morphological characters.[11] Dividing by the common functions of body and soul—by things like locomotion—may have the effect of splitting a kind, such as ants and glow-worms (642b33). Depending on the stage of life, ants can be either walkers or fliers. However, it is a question whether such a difference constitutes a difference in kind. Aristotle seems to be highlighting a serious problem; dichotomy cannot account for common functions, understood either as common between body and soul or common across different kinds.

Similarly, differentiating things by being either domesticated or wild will have the effect of separating things like wild dogs from domesticated dogs (643a5). The issue of domestication points to the city, as Aristotle indicates by the suggestion that there can be wild and tame human beings (643b6). Domestication, whether in the form of farming or ranching, is an art practiced in the city. Is dividing by wild and tame dividing along a natural joint or one borrowed from the arts? Perhaps it is important to note that Aristotle here recognizes a form of selection, artificial selection, when appealing to the different "kinds" that result from domestication.

It is at the point when the domestication offered by the city is introduced that Aristotle first hints at what looks like the alternative to dichotomist division, which is to mark off the groups by many differentiae, as is done popularly (643b11).[12] Insofar as we use the popular names in initiating our study of animal kinds (642b14, 643b11), we are starting from the opinions found in the city. The opinion about what constitutes the group "bird" furnishes a ready-made collection; each of these

[10]The Greek is ambiguous here and could be rendered (i) the actions common to body and soul; (ii) the common actions of the body and the common actions of the soul (Balme 1987b, p. 76).

[11]While the Greek is ambiguous, Aristotle's examples of animal locomotion make it clear that he means those attributes that are common to both body and soul, in other words those things that illustrate the nature of the ensouled animal. Therefore, I am not convinced by Balme (1987b) or by Lennox (1987) that Aristotle is referring here to the common actions of the body and/or those common to the soul, e.g. those shared by many animals. While this makes sense in so far as the dichotomists do not adequately understand the same and other and how these relate to commonness, the examples here do not support this position. I will argue that the common functions (*tois koinois ergois*) do point to attributes, to the actions and passions (*praxeis kai pathê*) of the animal (see below).

[12]Balme (1987c) does not discuss the connection that dividing by many differentiae has with the process behind the popular designations; he only sees how the process is discussed in the context of Book I.4–5. This, I would suggest, differentiates my attempt to interpret the *PA* and the flow of the argument, so that I believe it is necessary to examine this criticism of dichotomist division in light of the concern with opinion or popularly designated terms.

groups has had its boundary marked off and constitutes a collection as much as, or more than a division. Aristotle seems to be presenting the practice behind the popular designation as an alternative to dichotomous division. Aristotle's suggestion of the popular determination as an alternative may be an attempt to highlight the need for synthesis—the designation of the *genos* like "birds"—and the analytic determination of *eidos*. Of course, opinion may make a split of something that the natural philosopher collects in a sense, or recognizes as a whole. If opinion expressed in popular designation provides a starting point, it is not an adequate conclusion of analysis.

3.6 The Discontinuity of Dichotomous Division; Or the Part/Whole Relationship in Discourse

We continue to have to ask ourselves, what do we reach when the process of dichotomous division, or the process of differentiation that Aristotle seems to be advocating for that matter, has come to an end? Do we reach something that is analogous to the popular distinction that identifies birds and fishes? We are told that the last differentia will be the final species (643b17; *teleutaion eidos*): Is the final *eidos* a group like "fish" or is it something like a particular form of the group, like bream?

Despite what Aristotle had claimed earlier about the last differentia being the only valid differentia (642b8), the dichotomist considers the vertical elements of any given branching tree. This is why Aristotle can say that the compounded differentia—that which is out of plaiting (*ek sumplokês*)—is a collection of the vertical elements (643b30) (e.g. the collection of "A", "C", and "G" in Fig. 3.1). If some collection of the various elements from the various lines of differentiation is required, one is confronted with the problem of how to proceed in this activity. If, when dividing dichotomously, one does not take the differentiae of the differentia, all that remains will be a jumble of descriptions that have to be assembled with connectives, similar to the way speeches can attain a spurious kind of unity with connectives like "and" (643b18). If one divides the differentia "feathered" by "tame" and "wild," then one is not taking the differentia of the differentia, but dividing in a discontinuous manner. Taking the differentiae of the differentia shows the continuity or wholeness of that which is being divided (643b35). The idea that the divisions should have the integrity of a whole is a novel turn in the argument. Just as a speech has a certain wholeness that a jumble of fragments joined by connectives does not, the division of living things should also have a kind of wholeness that is not guaranteed by, much less a concern of, dichotomous division.

In Plato's *Phaedrus* Socrates suggests that every discourse be structured like a living creature, composed in such a way as to be consistent with the whole (264a). According to the *Phaedrus* account, the complement to collection is division characterized by the ability to divide into kinds according to natural articulations without shattering the unity of a natural part, like a clumsy butcher might do (265e). As a variation on the theme in the *Phaedrus*, Aristotle takes the dichotomous divisions that attempt to discover the whatness of living things and suggests that those divisions strive for a kind of wholeness that is not present in the dichotomous division of the Academy.[13] As it stands, however, dichotomous division does not meet the standard that is achieved when one considers a whole of parts. The *Phaedrus* indicates that there is some analogy—through the application of part-to-whole structure—between living things, discourse and division of kinds. Animals are wholes of parts. Any attempt to divide animals into kinds must also be organized as a whole of parts. Now we see more clearly why there is an ambiguity between "part" and "species" or "kind." If division, dichotomous or other, is a type of discourse aiming at the determination of kinds, then it will be arranged with an eye toward the parts that make up the whole. But simply because the discourse, the division, should hang together as parts of a whole, should the subject matter at hand, in this case living things, also be understood as parts of a whole? In other words, we must ask whether the kinds that are determined by our divisions should be viewed as parts of a larger animal or organic whole. An examination along these lines will have serious implications in determining how Aristotle understood teleology in nature.

3.7 Plaiting and Dividing

The method of dichotomous division makes it seem as if the final differentia is the only one (643b37). However, far from being superfluous, as Aristotle initially indicates (642b5), vertical elements of any given line may need to be collected in order to determine a final differentia—operating here is a distinction between the differentia arrived at lastly (*aphikneitai*) and the "final" (*teleutaian*) differentia. Not only do the vertical elements have to be incorporated, but also, like pieces of thread, different lines of division have to be plaited together in order to formulate the final differentia, as opposed to the last differentia (644a2). The collection or plaiting together that follows division moves vertically within a given line and horizontally between lines. The final differentia, which might turn out to be the *eidos*, may be the collection of a number of differentiae, vertical elements, along with different lines of division woven together. Even if the dichotomists allow for compounded differentia—those made of a combination of the vertical descriptions along a tree (643b30)— the need to move between disparate lines in a horizontal manner

[13] Pierre Pellegrine makes a very interesting remark in light of the present discussion when he suggests that "like living organisms, great works adapt themselves to new environments in bearing fruit inconceivable to their own authors" (pp. 120–121). Apparently without thinking of the *Phaedrus* discussion, Pellegrine, it seems, recognizes how works can, in a certain sense, take on a life of their own.

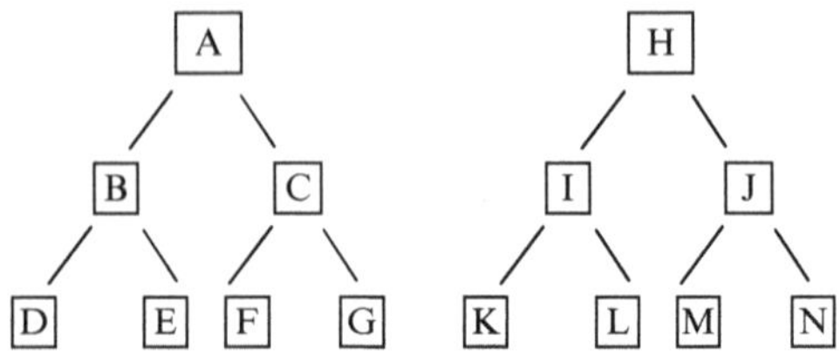

Fig. 3.8 A representation of different lines in dichotomous division, whose elements need to be collected in the final analysis

(e.g. F and C and A and L and I and H in Fig. 3.8) may amount to transcending the bounds, assumptions and principles of dichotomous division altogether.

Aristotle uses the human being as a subject of analysis to illustrate the necessity of moving horizontally between lines of dichotomous division. If man were merely a fissiped, one could compile the vertical elements of a division that is continuous—the result of taking differentiae of differentia—and be left with humans as footed-bipedal-fissipeds. The fact that human beings are not of this simple category demonstrates the need for the horizontal movement between lines of division, in order to capture the final differentia (644a6).

Determining that the final differentia must be composed of many differentiae not under one division (644a7) has several consequences. We have the possibility now of a double movement, vertical and horizontal, as elements of our divisions are collected and plaited together. Privation or nonbeing can now be dealt with or accommodated in a way not offered by dividing dichotomously (643b23). We have been given a possible relationship between the act of dividing kinds as a form of discourse, and the organisms that are the subject matter of such a pursuit. We also glimpse the idea that a part-to-whole analysis might apply to both the inquiry and the subject matter.

3.8 Swimmers and Fliers Revisited

Now that the argument has seemed to dispatch dichotomous division, we can return our attention to a problem that was previously glossed over when Aristotle appealed to the process behind popular designations as an alternative to dichotomy. Despite faulting the dichotomists for mixing birds and fishes (642b10), Aristotle returns to the problem to wonder why there is not a recognition of some kinship between fliers and swimmers (644a12). For example, opinion does not recognize the similarity in movement between a large ray and a raptor, or see how the movement of a wrasse's pectoral fins resembles the beating of wings. In bringing up the problem of fliers and swimmers again, Aristotle examines opinion, that which provides the basis of separation between birds and fishes (642b14). While it looks as if Aristotle defends the correctness of this opinion, he takes this opportunity to point out that birds and fishes, as well as all animals, have certain passions (*pathê*) that are common across kinds (644a15). Thus, notions like same and other are necessary for determining kinds (643a12 ff.). We have noted that dichotomous division seems to have a difficult time appreciating and accommodating sameness and otherness (e.g. 642b33, 643a13). Related to this is the problem of common differentiae (643a8). If one cannot see how some different kinds can be in some sense the same, whether in movement or the morphology of a part—I am thinking here of the discussion of the similarity between the pig's snout and the beaks of certain birds (662b12) that allows for a certain sameness in activity—one cannot examine things like the common attributes of the organic world (e.g. 639a17). On the other hand, we can note that the practitioners of dichotomous division want to get from *genê* to something more ultimate and particular (642b5), in contrast to the process of popular designation (cf. 644b3).

Looking at division embedded in opinion between fishes and birds provides an opportunity to refine what we mean when employing the notion of otherness. Otherness within a recognized kind is indicated by the more and the less (*mallon kai hetton*), while otherness between kinds is indicated by analogy (644a17).[14] Feathers of birds, while in some sense being the same, can differ by the more or the less (644a20): feathers on one species of the kind bird can be longer or shorter than the feathers of another. The difference between fish scales and bird feathers, on the other hand, means that any likeness between them can only be a matter of analogy (644a21). We have suggested that differentiation (*diaphorein*) involves not only division, but also some collection or plaiting together. Similarly, sameness and otherness are refined or replaced by the notions of the more and the less and analogy, which serves to indicate similarity and difference (644a23). The more and the less and analogy are a result of the plaiting together of sameness and otherness: Scales and feathers are obviously other, but may be similar in terms of the purpose each serves for the organism and so analogous.

Leaving aside the question as to whether dichotomous division and popular designation even aim at the same thing (cf. 642b5 and 644b3), Aristotle first addresses dichotomous division using the process of popular designation as a foil, then examines the latter more carefully once it appears the former is no longer a concern. He examines these two alternatives because of the way each deals with notions like sameness and otherness, which become reformulated into the more and the less and analogy. These notions are very instructive when brought to the study of the parts and lives of animals. The discussion of dichotomous division and the process behind the formation of popularly designated groups has a direct bearing on how we are to understand whether to proceed by beings or by common attributes (639a16). The connection between the earliest problems of the *PA* and the discussion of Book I.2–4 hangs on the better understanding of sameness and otherness offered

[14]On the use of analogy in biology, Pellegrin (1986) says that it is not "to set apart natural families of living things as to relate one group of animals to another by some point of reference, and ultimately to relate all living things to one unique being, taken as a model of intelligibility, man" (p. 91). He notes later that "analogy directly implies a combination of the same and other: in the case of wing/fin, for example, the otherness is immediately perceptible, for wings are not fins; their similarity flows from their both belonging to the *genos* 'locomotive organ." (p. 127).

by the more and the less and analogy. Only at this point can the argument return to the question of whether the inquiry into nature should primarily examine beings (*ousiai*) or attributes that are common to different kinds.

3.9 Beings and Immediate Forms

Substantial beings (*ousiai*), Aristotle tells us, are immediate forms (*eschata eidê*) (644a24). Aristotle formulates what he means by *ousia* as that which is indivisible in form (*to tô eidei atomon*) (644a30), undifferentiated in *eidos* (*ta to eidos adiaphora*).[15] Aristotle gives the examples of Socrates and Coriscus to illustrate what he means. The immediate *eidos*—for example, Socrates—is an *ousia*. In the *Metaphysics*, Aristotle argues that the universal is not a being (*ousia*); "for none of what is common signifies a this but only a such, and a being is a this" (1003a8–10). This issue lies behind the question, raised in the *PA*, of whether or not to start with each being separately or to begin with the common attributes. Socrates is a this, and hence, Aristotle claims here, an *ousia*, but we must wonder whether the inquiry into nature should begin with a being like Socrates or with attributes that are common.

With the suggestion that immediate forms are *ousiai*, which are indivisible or atomic (*atomon*), we are left to ponder the way in which we should proceed in our investigation of the parts of animals or with the *historia peri phusis* in general. We could restrict any part of our investigation to the *ousia*, the indivisibles (644a 28). If we were to do this, we would take up cranes—one of the elements that makes up the *genos* "birds"—and outline the attributes that the *eidos* "crane" contains. Such a procedure will, as Aristotle reminds us, necessitate our having to repeat ourselves—certain activities and parts will be common across *eidê* of a given *genos* and possibly common across *genê*.

Aristotle provides us here with two additional examples of that which he considers particular and indivisible (*hekaston kai atomôn*) (644a32): man and bird. He goes on to claim that birds are a *genos* that contain forms (*eidê*) (644a33). Unlike Socrates and Corsicus, who were given as examples of that which cannot be further differentiated in form (*ta to eidos adiaphora*), the category "birds" would seem to include a plurality of forms or *eidê*. How can something that allows breakage into *eidê* be said to be particular and indivisible? Even if the *eidê* of the *genos* "birds" were thought of as parts in a whole, it would be paradoxical to consider that genus particular and indivisible.

In contrast with birds and fishes, there is no recognized *eidê* further dividing human beings. However, in the case of human beings, Aristotle suggests that we

[15]We might have been left with the impression that everything associated with dichotomous division was rejected, but as this formulation indicates, Aristotle is concerned with getting at the indivisibles, which was a goal of dichotomous division (643a7), but not of the process characterizing popular designations. Dividing ceases with the un-differentiateable (*adiaphora*), the immediate *eidê*.

speak not by *eidê* but by particulars (644b6). In this context, Aristotle is willing to use the popular designations that exist regarding non-human animals as a starting point, but not the political or cultural distinctions among human beings. Perhaps because this discussion is embedded in a biological and not a political work, Aristotle does not have recourse to the division of humans into, say, Greek and barbarian (*Statesman* 262d, 263c); in fact, even the distinction that was drawn between domesticated and wild human beings (643b6) is not brought up at this point (cf. *Statesman* 264a).

Certain affections (*pathê*) will be common to many *eidê* (644a33). Consequently, it appears as if our study of the parts of animals must include passions or affections that are the result of a creature's being ensouled. Aristotle had criticized Democritus for not distinguishing between the living and the dead: he did not recognize soul—or function as Aristotle had couched the discussion (640b30 ff.). In bringing in the passions—that which moves the soul—Aristotle reaffirms the notion that we will be investigating living things. It looks as if the criticisms leveled against Democritus could also apply to the conventional names assigned to certain kinds.[16] Popular usage is based on recognition of the shape or form of the organism (644b7); it is not clear that it takes into account any notion of function or soul. The process of popular designation of *genê* does not encapsulate everything Aristotle wishes to investigate: it recognizes, for example, something like Aristotle's notion of the more and less, but neglects a corresponding concern for analogy (644b13–15), which is a sign of its neglect of function. That popular usage abstracts from soul is indicated by Aristotle when he suggests that *genê* like fish and cephalopods are distinguished by the bodily affections (*somatikois pathesin*, 644b12). We may start with popular usage, consequently starting with bodily form or shape, but it is clear that Aristotle finds this inadequate for analyzing ensouled bodies. Just as the inquiry will move from the common to the particular, it will move from shape or body alone to the ensouled, to animals as functioning beings in an environment.

3.10 Bringing the Gods Down Into the Inquiry

Nature as a whole appears to divide between those things which are subject to becoming and corruption and those things which are not: "Of the substantial beings (*ousiai*) naturally united, some are ungenerated (*agenêtous*) and imperishable for

---

[16]Balme (1987b) does not recognize that the *PA* contains a criticism of the process of popular designation; he resorts, therefore, to a discussion in the *HA* to see that Aristotle's emphasis on morphology is to be tempered in his own method (p. 79). On the other hand, in arguing against the common view of the *HA*, Balme makes an important general point that is appropriate here: "The belief that there must be a classification in the background rests on the assumption that Aristotle, like every good pre-evolutionary zoologist, put systematics first in zoology and morphology first in systematics. But better sense can be got out of the evidence if this dual assumption is removed" (p. 79). Balme's general argument, that Aristotle did not engage in any attempt at systematics for the sake of systematics, but rather in a search for causes, is important to keep in mind.

the whole of ages, but others are subject to becoming (*geneseos*) and perishing" (644b21–23). Since we are mortal, much of what we have access to is subject to becoming (644b29). The ungenerated beings are said to be honored and divine (644b25). In the beginning, different studies were ranked in light of the honor in which they were held; now, we are presented with the objects of subjects, the natural beings, in the light of honor. Corresponding to the division between generated and ungenerated beings within nature is a division among sciences or inquiries: there can be a philosophy (*philosophia*) of the ungenerated divine beings (645a4) or of those beings which are subject to becoming. In the first chapter, Aristotle had claimed that certain demonstrations can be appraised in light of something other than truth or falsehood (639a14). The sciences ranked on the basis of the distinction between generated and ungenerated things of nature are evaluated in light of some value or use and pleasure. Thus, the inquiry into nature can be pleasing and useful, despite what was implied in the opening chapter of *PA*. In fact, the study of animals will produce immeasurable pleasure (*amêchanous hêdonas*) to those who are by nature philosophically inclined (645a9). The pleasure gained by such an investigation does not appear to depend on the honor of its objects, as in the study of the ungenerated, divine things (644b33). It turns out that we can be pleased by a study of things not considered honorable.

The study of animals and other things subject to becoming and perishing will also have a certain relevance to our life and our understanding of what we are (645a5, 645a28). In recounting his intellectual biography, Socrates criticizes the Anaxagorean account of mind, suggesting that he would give the cause of Socrates' sitting in prison in terms of his bones and sinews or of his conversing in terms of the operation of voice and air and hearing (*Phaedo* 97d). Despite his surprisingly detailed description of the mechanical processes (see Burger 1984, p. 142), Socrates turned away from describing causes in this way. Aristotle admits that "if anyone has thought the study of the other animals valueless, he should think the same about himself; for one cannot without considerable distaste view the parts that compose the human kind (*tôn anthrôpôn genos*), such as blood, flesh, bones, veins, and these parts" (64a27). He implies that we really need to know the blood and guts of the matter, if we are to understand humans as part of the animated world. Socrates turned away from a direct examination of the beings themselves in favor of examining their representations in speeches. Aristotle seems to be rejecting this turn, at least for the study of nature (cf. 645a11). In all natural things (*phusikois*), Aristotle tells us, there is something wonderful (*thaumaston*); and as we are informed elsewhere, a state of wonder is the beginning of philosophy (*Meta*. 982b12–19; *Theaetetus* 155d, *Phaedo* 97a). If Aristotle had in mind the Socratic move from the first way to the second sailing, he should be concerned especially with the role of teleological and mechanistic accounts (*Phaedo* 97d ff., cf. *PA* 642a2).

In the current context, Aristotle makes reference only to the final cause, the for-something's-sake (645a24), and not to both this and the necessary, as he had done previously (e.g. 642a2). It is striking that Aristotle leaves out the latter cause, the necessary, in what seems to be an allusion to the *Phaedo* account. If the Socratic turn can be described as a turn from the Aristophanic Socrates who has his head in the heavens, or at least in the clouds, to a Platonic Socrates, who brings philosophy down from the heavens to the earth, Aristotle seems to want to present this study of the parts of animals in somewhat similar terms. Perhaps ironically, Aristotle uses Heraclitus to illustrate the movement from the heavens to the earth: he relates how Heraclitus was said to have beckoned hesitant visitors into the kitchen saying "there are gods here too" (645a22). This, as Aristotle presents it, is how we are to think of the inquiry into animals, for in them is something natural and beautiful.[17] Where is the divine to be found in the realm of nature? Among the beings that are ungenerated as stated at the beginning of the chapter (644b26) or among the beings subject to becoming and corruption as is implied here (645a22)? At important points in the rest of his argument, Aristotle will suggest that humans might have something of the divine in them (e.g. 656a13). This indicates a bringing of philosophy down from the clouds.

3.11 A New Turn in Division

The argument of the *PA* begins with the division of knowledge into what appear to be sciences (639a) and then moves, with the discussion of dichotomous division, to the division of animals into kinds. As the end of Book I approaches, the focus moves to the division of attributes, instead of the particular animals that possess them: "It is necessary (*anagkaion*) first to divide-off (*dielein*) in relation to each particular *genos* (*prei hekaston genos*) the attributes (*ta sumbebêkota*) that exist for all animals and after this try to divide-off (*dielein*) their causes" (645b1–4). Division is now going to focus on separating the attributes from the animals and then on dividing off the causes of these attributes. This statement of how the inquiry is to proceed sheds light on a prior problem: first identify attributes common to all animals, we are now told, and then attempt to determine the causes of these attributes (645b). This seems to be the answer to the question about whether to take each being singly or proceed by first recognizing common attributes (639a16). Why does the apparent answer come only now?[18] Perhaps Aristotle felt it necessary to wrench the activity of division from the dichotomists, who wanted to focus on arranging animals, in order to free it up for dividing-off attributes and causes. The focus in our search for causes will be on the attributes instead of the animals themselves that possess such attributes.

---

[17]We can note that Heraclitus' reference to the gods or the divine is replaced by the beautiful.

[18]Other commentators do not raise this question about the order of the argument. They do attempt to connect these points in the *PA* with the *Apo*. II. 14–18 (e.g. Lennox 1987, p. 114 ff), but they are not concerned with how the argument moves from the positing of the question to the answer and what comes in between. Instead they view *PA* I as "a string of papers" (Gotthelf 1987, Balme 1987b). Lennox (2010) has recently argued that there is a narrative integrity to *PA* I. I have attempted to give a plausible account of this movement through Book I and will attempt to make sense of the movement of the argument through the rest of the *PA*.

Division of the attributes works for our study of animal life because "many [attributes] belong to many animals, some simply (*haplas*) (like feet, feathers, scales, and affections (*pathê*) too in the same way), but others analogously" (645b7). Thus, one can separate the attributes from the things—in this case, animals—that possess them. While feathers and scales have been used to illustrate the notion of analogy (644a21), here we have feathers and scales included in attributes that belong simply to many animals, in contrast to those identified as analogous. We are thus forced to puzzle over what analogy actually entails. This problem regarding the similarity and difference of scales and feathers is made even more pronounced by the recurring reference to the question of water animals (fishes) and fliers (birds) (642a10, 644a13). Both feathers and scales are parts for protection in the way that fins and wings are similar in belonging to the "*genos* 'locomotive organ.'"[19] However, it is difficult to consider fur or skin as analogous to feather and scales. Animals have common attributes like protective coverings or locomotive organs that show a great range of diversity when studied at the level of particular organisms. However, it is in studying the movement of particular fish or birds that one recognizes the analogous relationship that scales and feathers have. The need for locomotion or protective covering or reproduction seems to be a problem or question that is answered in many different ways by the particular organisms.

The examples Aristotle uses of analogy in this particular context include lungs and whatever is analogous to lungs—an organ which obviously does not spring to mind, but which Aristotle suggests later is something like the gills—and blood and whatever has the capacity (*dunamin*) of blood in the bloodless animals (645b6). Some capacity determines whether parts are analogous. The same, whether simply or through analogy, is present in many, so that to speak of each of the particulars (*hekasta*)—where particular here refers to the attributes of animals—is to have to repeat oneself (645b11–13; cf. 644a26, 639a25).

The method by which we are to proceed now is to divide off the attributes that belong to each kind (*hekaston genos*) and then divide off the causes (645b). However, this understanding of the method changes from the perspective that sees a given part as an instrument (*organon*) for-the-sake-of (*heneka*) something—for-the-sake-of an activity (*praxis*)—so that it appears (*phaneron*) that the body as a whole (*to sunolon soma*) is for-the-sake-of some many-parted (*polumerrous*) activity (645b14–17; cf. 642a12). The whole, that is, the body, is for-the-sake-of some many-parted activity. This formulation suggests that one whole, the body, is for the sake of another whole, some activity with many parts. The effort to divide the attributes from the animals would then result in two wholes of parts, the many-parted body and the many-parted activity of the animal. Parts can refer, then, to those of the body or to those of the multi-parted activity. A saw, to use Aristotle's example, is a whole, but so too is sawing, which is useful (*chrêsis*); in fact, the useful characterizes the activity as a whole.

[19]On this, see Pellegrin 1986, p. 127.

This picture is deepened when Aristotle introduces functions and soul: "The body is for-the-sake-of the soul, and the parts of the function (*ta moria tôn ergon*) to which each [animal] is naturally-adapted" (*pephuken*) (645b19).[20] Function has been introduced together with soul and the idea that an animal's activity makes up a whole of parts. As a consequence, the method gets re-worked (cf. 639a16, 645b): the "attributes that belong to all animals" (645b2) becomes the activities (*tas praxeis*) that are common (*tas te koinas*), those that are according to kind (*kata genos*), and those that are peculiar, according to form (*tas kata eidos*) (645b21; see Fig. 1.11).[21] What was two—the common or general (*koinê kata genos*) and the particular (*tôn idiôn*) (639b5)—has become three levels of complexity—the common, the generic according to kind (*genos*), and that according to form (*eidos*), which replaces what was originally described as the particular.[22] All three levels address what is in common, what is in some sense the same either through analogy, degree (presumably in terms of the more and the less), or by the fact that it is an *eidos*, that which has no differentia in its general logos (*katholou logon*) (645b25). The human being is used as an example of this last level, the level of the *eidos*, just as it had been used to illustrate the particular (*hekaston*) (644b7). But how we are to identify what is common or same within this last level remains unclear.

After the activities—those that are common, according to kind, or according to form—have been articulated, we must examine activities that are for-the-sake-of others. Doing so will help to identify the activities—which are presumably ends (645b27–30)—that are parts of the life activity of the animal that is a whole. It is also important to see how some of these activities exist from necessity (645b33). Determining the relation between the necessary and that for-the-sake-of-which (642a2) will enable us to articulate the causes (646a3).

Since the argument has left us with two wholes of parts—the body and the activities—the attributes (*ta sumbebêkota*) can be split along these two lines. On the one hand, there are the affections and activities (*pathê kai praxeis*), which include genesis, growth, coition, waking, sleep, locomotion and others; on the other hand, there are things like the nose, eye and the face as a whole, each of which is called a member (*melos*). It is important to note that neither of these sets of attributes is referred to as "parts" here. The task of the *PA* must involve showing how all of these, the passions and activities as well as the so-called members, are parts of the animal in motion, the animated animal.

[20]Alternatively, the line can be translated as "consequently, the body too is in a way for the sake of the soul, and the parts are for the sake of the functions in relation to which each has naturally grown."

[21]At this point, it seems to be difficult to accept the claim made by Balme that the treatises, like the *PA*, are less concerned with animal activities than the latter books of the *HA* (1987a, p. 14).

[22]While many commentators recognize that much of the argument of the *PA* takes advantage of the double explanations afforded by appeals to both final and necessary causes, they do not attempt to make sense of the reformulation of the method that is offered here.

Chapter 4
The Examination of the Animate in Light of the Inanimate: Or, the Argument for the Autonomy of the Zoological Inquiry

Abstract It should not be surprising that Aristotle concerned himself with questions of the mechanics of material and functional explanations. *PA* II beautifully illustrates this two-parted investigation. Aristotle's *PA* II falls into two parts, marked by the announcement of a "new beginning" in II.10. The best way to understand the move from the first to the second half is to note the status of the animate whole, the organism, and its relation to the inanimate.[1]

At the beginning of Book II, Aristotle offers a suggestion as to how one might differentiate the present inquiry, the *PA*, from his *History of Animals* (*HA*). The *HA*, we are told, presents detailed descriptions of various organisms, cataloging their morphology, physiology and the peculiar facts of their life histories. These discussions present the "what" and "how many" of the parts. The *PA*, in contrast,

---

[1]Where is biology's place in the web of human cognitive capacities? Rosenberg (1985) claims that "whether and how biology differs from the other natural sciences... is the most prominent, obvious, frequently posed, and controversial issue the philosophy of biology faces" (p. 13). More recent evidence of the persistence of the "whether and how biology differs" question is provided by Marc Lange's reply to positions Rosenberg has recently taken up on the issue of general laws. Rosenberg (2001) argues that "general laws must be timeless truths" and "no such laws are attainable in biology" (p. 141). Such a commitment inclines Rosenberg to view functional explanations in biology as in need of "completion and correction by macromolecular explanations, which apply the principles of natural selection, along with the laws of chemistry and physics, to initial macromolecular and environmental conditions" according to Lange (pp. 94–95). In contrast, Lange has a strong argument in favor of the autonomy of functional explanations by defending the claim that these reflect the "fact that the range of stability exhibited by 'The S is T' generalizations extend in some respects beyond the range of stability exhibited in physics by the laws grounding the sorts of explanations that Rosenberg favors" (p. 108).The real disagreement between Rosenberg and Lange makes certain that that "most controversial issue" in the philosophy of biology is alive and well. As will be shown in the body of this chapter, Aristotle's argument can be seen to dialectically move between the mechanics of material—which in contemporary debate looks like the laws of chemistry and physics applied to initial macromolecular conditions—and functional explanations. While this can be seen in many places in the Aristotelian corpus, the second book of the *Parts of Animals*, *PA* II, best exemplifies the paradoxical fruitfulness of attempting to incorporate both poles into the account.

J.A. Tipton, *Philosophical Biology in Aristotle's Parts of Animals*, Studies in History and Philosophy of Science 26, DOI 10.1007/978-3-319-01421-0_4,

hopes to uncover the causes by which organisms are what they are. Aristotle indicates that the *HA* gives evidence for the way that animals are; the goal of the *PA* is to gain some knowledge of the why of things, some knowledge of cause.[2] That the discussion of the causes will revolve mainly around the final and the necessary was anticipated in the first chapter of the *PA* (642a2 ff.). This expansion of the themes first touched upon in *PA* I is consistent with Lennox's (2001) suggestion that "a strong case can be made that Books II–IV reflect the philosophical standards that are articulated in Book I" (p. 179; see also p. 181). A very good illustration of this principle is to be found in *PA* II; as we hear at the end of *PA* I (645b21 ff.), the investigation will frame things in terms of the attributes which are common, those according to *genos,* and those that are particular (*idion*), according to *eidos*. This threefold structure, I want to suggest, will help to illuminate the teleological-necessary framework of the inquiry. The discussion of common, generic and more specialized functions comes to the fore in the second half of Book II, and helps distinguish it from the orientation of the argument of the first half of Book II; this, then, is also a way in which to understand the two halves of *PA* II.

Within the framework of the final and the necessary, Aristotle incorporates details about matter and the material; the living being is examined as a composite of matter or stuff. This makes manifest a problem regarding two apparently incommensurable kinds of matter, animate and inanimate. In describing the genesis of animals and their parts, Aristotle presents it as if one can retreat all the way back to the elements, back to the inanimate. This amounts to giving an account of heterogeneity—the heterogeneity that is manifest when we view the living world—in terms of the homogeneous elements. This would be to attempt to understand the complex in terms of the simple. The movement of the argument of Book II attempts to ground the diverse and varied actions and movements of animals in the varied potencies (*dunameis*) of the elements, or of the material: an attempt is made to explain the diversity of the animate by means of an appeal to the diversity of the potencies of the elements. With the introduction of uniform and non-uniform matter in *PA* I, Aristotle refined the material account offered by his predecessors. *PA* II continues to refine our understanding of matter, for instance, by showing the potencies behind material. The relationship between the uniform and non-uniform is explored further in *PA* II. In addition, the notion of the simple and complex, which does not correspond exclusively to the uniform and non-uniform, is brought into the discussion. This then, as we shall see, is a way to make what is generally considered homogenous stuff more heterogeneous in power. Evidence for the incompleteness or partial character of this argument that begins with the elements emerges when Aristotle claims that the natures of the elements are *nearly* the causes of animate phenomena (648b7).

---

[2]Actually, the reference to "*ta historia*" is not at all straightforward. It seems to refer to what we have as the *History of Animals*, but could also be a reference to an activity in which we are engaged (e.g. *PA* 639a12). The ambiguity as to what this formulation points to would then raise the question of the distinction between *historia* and causal explanations. Thus, it seems to me as if Lennox's (2001) translation of "*ta historia*" as "the enquiries," as opposed to the *History of Animals*, strikes me as quite sound.

The discussion of necessity in *PA* I might also illuminate the argument of *PA* II. It looks as if the emphasis on the elemental is to ground the argument in terms of simple necessity in that the elements could be understood as eternal; predecessors such as Thales were attempting to find some abiding substantial being (*Meta.* 983b10). In its emphasis on nutrition and feeding, the second half of *PA* II could be understood in terms of conditional necessity and the third form of necessity.

One might argue that the relationship between non-living matter and the living is non-problematic for Aristotle; non-living matter becomes living when it is integrated into the body of a plant or animal by its *eidos*. According to this account, different *eidê* require different sorts of matter to be instantiated; each kind has a material peculiar to it (*Meta.* 1044a18–20).

While this general picture of an *eidos* shaping non-living, inanimate material seems to account for most animate genesis, examples such as spontaneous generation complicate such an explanation. In some instances, Aristotle suggests, matter determines, to some significant extent, the form. So, he can claim that "all the testacea (*ostrakôdê*) arise by spontaneous generation (*gignetai automata*) in mud, though they exhibit differences according as the mud differs: in slimy mud oysters grow, in sandy mud cockles and the others mentioned" (*HA* 547b18). In a similar vein, Aristotle suggests that when seeds are put in a foreign environment, the material character of the locale will affect the form (GA 738b35). Place or habitat produces differences in shape or form in large part because of the material found in that locale. Such considerations demonstrate how the relationship between living and non-living matter is more complicated than it might first appear. The material or inanimate constituents of the animate whole need to be carefully analyzed in order to better understand biological phenomena even if, as will be shown, the account cannot move smoothly between the animate and inanimate.

Aristotle moves between an emphasis on a material account in the first half of his analysis (II.1–9) to the complementary emphasis on the functioning animate whole in the latter half of Book II (10–17).

4.1 Homogeneous to Nonhomogeneous: Final Cause

Aristotle distinguishes three kinds of composition (*sunthesis*, 646a12)—(1) the elements (e.g. water), (2) the uniform (e.g. blood) and (3) the non-uniform (e.g. hands)—one of which issues into another (see Fig. 4.1).[3] Any potential problem

[3]Lennox (2001) is surely right when he suggests that the discussion is more complicated than it might first appear in contrasting "simple and uniform . . . composite and non-uniform" (671a1): Lennox notes that "the simple/composite distinction is based on the number of material constituents in an object—one indicating maximal simplicity. The uniform/non-uniform distinction—the Greek terms literally mean 'with like parts' and 'with unlike parts'—rests on whether or not a body is indefinitely divisible into like parts (cf. *HA* I.1, 486a5–8). It is thus possible for there to be simple uniform, simple non-uniform, composite uniform and composite non-uniform parts" (p. 182).

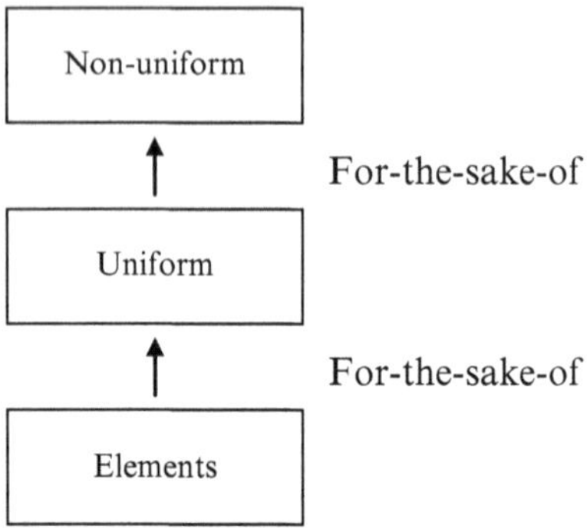

Fig. 4.1 The relationship between these three levels of composition as set up in the beginning of Book II. As we shall see, this relationship will be reversed in the course of the argument

of the relation between the animate and inanimate is suppressed in the hypotheses put forward in the beginning of the analysis. The elements, we are told, supply the matter (*hulê*) for composite bodies (*suntheton somaton*), while the uniform compositions form the material basis for the non-uniform (646a14 ff.). Each kind of composition has certain characteristics that contribute to the cause of the organism's parts and the structure of those parts. In moving from the inanimate to the animate, we are asked to consider the possibility that the becoming of the parts of animals begins with the non-living.[4] Insofar as the living animal incorporates non-living nutriment for the sake of its maintenance, there does not seem to be a problem. Although it is a topic that will be considered at some length, nourishment does not ultimately provide the answer to the question of the relation between the animate and inanimate.[5]

The one thing conspicuously absent from this hypothesis regarding the genesis of animate things is an explanation of the composition of the body as a whole, the organismal whole; what should be the last stage moving from elements to uniform to non-uniform is left out of the account. In other words, we get to the point at which the composition of non-uniform parts is explained, but are not told how these non-uniform parts come together with the uniform to form the whole body. We get an account of the synthesis of everything but the whole. Such an omission could perhaps be explained by noting that Aristotle does refer to the animal body as an *organon* (645b15 ff.), thus as a kind of non-uniform part. Is the entire organismal

[4]Gotthelf (1987) formulates the question as follows: "Is the development of a living organism the result of a sum of actualizations of element-potentials, or is it primarily the actualization of a single potential for an organism of that form, a potential the actualization of which involves the actualization of element-potentials, but is not reducible to them?" (p. 212).

[5]The suppression of the split between the animate and the inanimate may be a product of the Pre-Socratic assumptions with which Aristotle begins in this context. He takes the Empedoclean elements, divorced from Empedocles' mechanism of clumping and commingling (e.g. DK fr. 23), and puts them to use in the lowest level of synthesis (646a12 ff.; Furth 1987, p. 23–24).

body to be thought of simply as a non-homogenous part? The body is said to be for the sake of a many-parted action or for the sake of the soul. Putting things in terms of elements, uniform and non-uniform parts allows us to temporarily suspend the question of soul. The attempt to move all the way from the soul-less elemental to the animate whole is left incomplete. The deficiency of such an account will be indicated further by the "new beginning" of II.10, with its emphasis, in part, on completed animals (*tois zôis tois teleiois*) (655b30). The limitation of the account that focuses on the *sunthesis* of the elements, uniform and non-uniform is thus indicated by the absence of the whole within it.[6]

Working up from the elements as parts leaves one short of the whole; while Aristotle is certainly aware of this incompleteness, there is a benefit in getting one's hands dirty, so to speak, in this type of an analysis of the animate world. If it is correct to suggest that, at some important level, the whole is the cause of partness of parts, an important feature in our causal analysis is neglected, if only temporarily. While such an account leaves out a discussion of the animate whole, Aristotle demonstrates that one can understand a great deal by means of such an analysis.

4.2 Genesis and the "For the Sake Of"

The composition of the non-uniform necessarily relies on the existence and character of the homogenous or uniform, while the uniform stands in a similar relation to the elements. The fact that the elements must exist prior to the uniform and non-uniform from a temporal or genetic perspective suggests to Aristotle that the elements are for-the-sake-of the uniform, while the uniform are for the sake of the non-uniform (646b5). In this context, "that which is for-the-sake-of" reflects a temporal sequence. It is also indicative of some necessary condition: the elements are a necessary pre-condition for the genesis of the uniform or homogenous. Similarly, the non-uniform requires, as a necessary condition, the uniform. The non-uniform parts are constructed out of one or more of the uniform parts, but not vice versa (646b33) (Fig. 4.1).

Aristotle says that the non-uniform parts have reached the end (*telos*) and limit (*peras*) in the process of *genesis* (646b7). Again, we note that there seems to be a missing stage insofar as we do not know whether to understand the whole body—that which is made up of various uniform and non-uniform parts—as a non-uniform part or the sum of all those parts.

---

[6]In his account of these things in Aristotle, Furth (1987, e.g. p. 37) does not recognize the absence of the organismal whole in the account of the first half of Book II as a result of the movement of the argument. Instead, he resorts to things said in *PA* I (p. 46) to account for the omission in the discussion.

Within a genetic framework, the framework that emphasizes the genesis of animate phenomena, the notion of "for-the-sake-of" has something to do with a temporal process, but also with a recognition of necessary conditions. The movement of the argument turns temporarily, in an anticipation of a discussion to come, from this to the question of function. The uniform parts are for-the-sake-of the non-uniform since functions and actions (*erga kai praxeis*) belong to the non-uniform (646b11, cf. 647a22); in this context, functions and actions are things that belong to non-uniform parts. The question of function now is to determine the relationship, one thing being for-the-sake-of another. From the perspective of function as well as from genesis, the uniform are for the sake of the non-uniform. Within the genetic account, that which should have been the last stage—the stage including the composition of the whole body—was conspicuously absent. This temporary shift in emphasis toward function makes explicit the functioning or action or movement of the whole. While the non-uniform parts were said to be the end in the previous argument (646b8), in the analysis that takes its bearings from functions and actions there is an explicit reference to whole animals (*tois zôis holois*) (646b17), which according to my interpretation is a crucial anticipation of the argument of the latter half of *PA* II. There is a sense in which the first chapter has within it the material for the argument of *PA* II as a whole. The animal as a whole comes into focus when examining the functions and actions of parts; thus, this particular discussion, in many important ways, contains the seeds of the turn that is taken in II.10. The dialectical nature of the argument makes it possible for the themes which emerge after the new beginning to make a brief appearance.

The concern with function and the animal whole also emerges, if only temporarily, when Aristotle suggest that the uniform and non-uniform will vary according to an animal's function and substantial being or *ousia* (648a14–16). The examples used to illustrate this—eyes and eyelids—are discussed at length in *PA* II.13–14. Because the actions and movements of whole animals are varied or, literally, polymorphic (*polymorphon*), the stuff that makes up the parts of the whole must, of necessity, possess dissimilar powers (*dunameis*) (646b15). The notion that there is a polymorphism of actions and movements is also an important anticipation of the "new beginning" of II.10 which will be discussed at length below. The polymorphism of action and motion picks up on the suggestion that the body is constituted for the sake a many-parted action (645b16–17). The polymorphism of actions and movements is initially correlated with the varied powers of the elemental parts; we are in a position to better appreciate Aristotle's appeal to function and actions of an animal. The argument moves from the variability of actions to the various powers at work behind the elements. There are different kinds of uniform substances and elements, since the homogenous is really heterogeneous in power. Aristotle suggests that the reason for this is final cause (646b27). The heterogeneity of functions and actions of living organisms suggests that the elements must also be heterogeneous: the living organism requires elements with dissimilar powers, such as a combination of hard and soft parts (646b18). Any revision in our thinking about the elemental is a direct result of our analysis of animate wholes as they act and function in the world.

4.3 Divide Between Instrumental and Sensitive, Overcome Through Touch

The introduction of function in this context allows for a new way to think about the relationship between parts; Aristotle distinguishes between instrumental (*organikon*) and sensitive (*aistheteron*) in animals (647a3 ff.).[7] This new move in the analysis does not completely jettison the previous discussion; the new division seems, rather, to map onto the previous division between the non-uniform and uniform parts: instrumental parts are non-uniform, while sensation takes place in the uniform (647a4). Does this mean that, just as the uniform was said to be for the sake of the non-uniform (646b7, 646b12), the sensitives are for the sake of the instrumental? The organism as a whole must unite the sensitive and instrumental.[8]

The identification of the perceptual parts with the uniform seems to be an expansion on an assertion by Aristotle's predecessors that each of the senses corresponds to one of the elements (647a12), so that one would be fire, another water, etc. In being linked with single kinds, single *genê* (647a 8), the sensitives are understood to be uniform. However, in this identification of the sensitives with the uniform, we lose the distinction between the uniform and the elemental that had been established; of course, this makes it possible, perhaps, to avoid for now the problem of accounting for the move between the inanimate elements and the animate. We are presented with a picture in which the assumed simplicity of sensible objects requires a corresponding simplicity in the sensitives, the organs responsible for sensation. The sensible objects are in a relationship to the sensitives as agents are to patients, or, as Aristotle claims, that which has the power to be affected—the sensitives—must be what the other—the sensible object—is actually.[9] The senses

[7]The instrumental parts were briefly introduced at 646b24. On this, Lennox (2001) notes that "the distinction between instrumental and perceptual parts is surprising, given Aristotle's use of eye and nostril alongside hand and arm as examples of instrumental parts at 646b13–14. Perhaps it reflects the association of instrumental parts with animal *activity*, while perception is a *passive* capacity, a *being affected by* an object of perception—colour in the case of vision, sound in the case of hearing, and so on" (p. 183). I will argue that the "surprising" split is undermined by Aristotle's discussion of touch. Two kinds (the perceptual and instrumental) can be distinguished at one level of the analysis while grouped together at another level. To anticipate, after the new beginning of II.10 the instrumental and sensitive designations get separated from the uniform/non-uniform distinctions; in other words, at a certain point we are no longer concerned whether a sensitive or perceptual part is uniform.

[8]In his piece "The Autonomy of Biology," Ernst Mayr (1996) suggests that "perhaps the most concise way to characterize the uniquely different nature of *organisms* is to describe them as *hierarchically organized systems, operating on the basis of historically acquired programs of information*, a definition that does not apply to any inert object" (p. 103). This is not to suggest that biological phenomena conflict with physio-chemical processes (Mayr, p. 104). Again, a similar position, I argue, can be seen in Aristotle.

[9]Cf. a similar passage in *Generation and Corruption* (I.7): "The active and the passive agents cannot be absolutely similar; for if so each object would move itself, and everything would be in perpetual motion. Neither can they be utterly dissimilar. For how can two such essentially distinct

are discrete and therefore do not mix; it is well said, Aristotle notes, that sensation takes place in the simple parts of the body (647a15), the parts that are not composed out of mixtures. Thus, they can be thought of as elements.

Touch presents certain difficulties (647a14 ff.). In its case, we seem to have an instance of an intermediary between the simple sensitive parts and the multifaceted instrumental parts; touch bridges the gap between simple and complex. The instrumental parts are combinations of dissimilar elements, dissimilar powers. For example, a non-uniform part may be composed of the fluid and the solid (646b11). Unlike sight, which is the most sensitive and discriminatory capacity, touch works through a series of opposites, such as hot and cold, fluid and solid (647a19). It is in its treatment of pairs of elements that touch stands halfway between the simple sensitives and the multifaceted non-uniform or instrumental parts.

Still, there is another way in which touch plays this intermediate role. By his choice of words, Aristotle indicates that the instrumental (*organikon*) parts, on the one hand, affect that with which they come into contact, they are meant to do some work. On the other hand, sensitives like sight, hearing, and smell, seem as if they affect nothing in their operations, there is no contact between sense and thing sensed. In touching, however, one senses and also affects; as Aristotle says, touch is the most corporeal sense (647a19). In this way touch stands between the working of the instrumental parts and the passivity of the other sensitive parts. Touch seems to be a means by which to overcome several dualities, something about which Aristotle will have more to say (656b35 ff.).

Reminding us of the necessary (647a23), we are told that there must be uniform parts because they are necessary for sensation and, especially important, there cannot be an animal without the power of sensation. Insofar as the soul is the form and substantial being, it will not be defined without its activity, which will include perception (*Meta.* 1035b15–22). Uniform parts are necessary for sensation, and sensation in turn is necessary for the animal to be an animal (647a23). This role of sensation is meant to complement the role of function. An animal and its parts must function in order to be an animal (640b30 ff.). It is not just that the account is to show how the necessary and final cause are intertwined: it is necessary for an animal to have sensation to be an animal; a functioning animal is also a complete or final animal.

The uniform sense organs and non-uniform instrumental parts have to be united somewhere in order for the animal to be a whole. If some part had the characteristics of the uniform and non-uniform, it could serve as a means of overcoming the split that manifests itself when we move from the uniform and non-uniform to sensitive and instrumental. The heart, or its analog, is meant to play this role: the heart can

things as whiteness and a line affect each other; but this is in virtue of their generic resemblance, both having color. So also a savor cannot affect a color, nor a color a savor; but one savor affects another savor, one color another color. The result then is that the passive and the active agents are generically one, but specifically distinct." It should be noted that this is similar to the kind of physiology of perception found in Plato's *Theaetetus* (151e ff.).

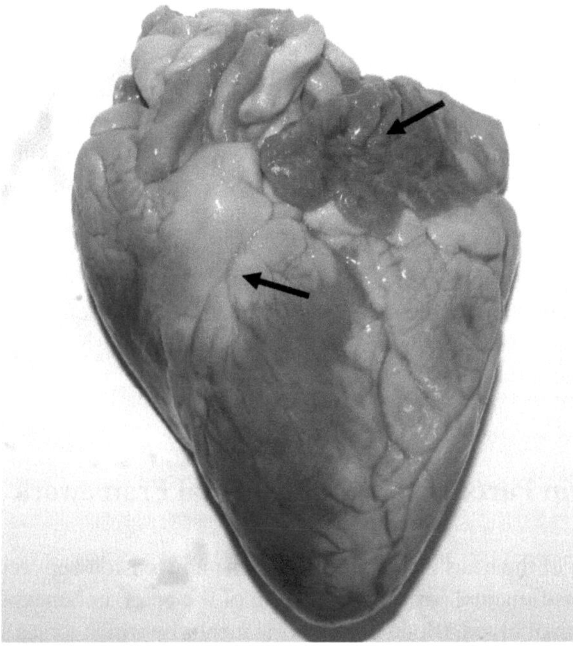

Fig. 4.2 Sheep's heart showing the structure to which Aristotle is alluding. The arrows point to one of two auricles and the line that indicates the internal division of the ventricles. Aristotle claims that the heart has sutures similar to that of the skull (667a7). He says that these sutures are lines of articulation in contrast to lines of conjunction. Normally, one says that the four chambered heart consists of two ventricles and two auricles or atria. While Aristotle must have been confronted with these four different structures, he argues that hearts such as the one above have three chambers. Students at St. John's College dissect sheep and beef hearts in order to understand William Harvey's *The Circulation of the Blood*

be divided up into uniform pieces, but also has a certain figure (*schêmatos*) or shape (*morphê*) indicative of the non-uniform (647a30).[10] The heart is both non-uniform (determinate) (See Fig. 4.2) and uniform (indeterminate).[11]

At this moment in the investigation, Aristotle pauses to make a general claim about the becoming of viscera. The viscera in general are formed like deposits of

[10]On things like the heart, Furth (1987) notes that "it may seem odd to us, but it seems we are being told that certain 'parts' are proto-structural, of an intermediate nature or at an interface between uniform and non-uniform: the same nature somehow doubling as stuff and structure" (pp. 34–35). The importance and role of such parts indicates to me the difficulty in overcoming certain divisions that must be overcome in a whole of parts.

[11]Or, as Lennox (2001) puts it, "the claim that the heart is both non-uniform and divisible into uniform parts rests on the fact that the visceral material is uniform throughout, while the heart has a number of features—walls, chambers, valves—that make it structurally complex" (p. 184).

mud left by a stream (647b). On such a model, we understand the coming to be of the viscera as we would the emergence of a sand bar in a stream, which is exactly the way in which Aristotle says his predecessors would account for such things (640b15 ff.). With the introduction of the heart, we have another way in which to understand the composition of the viscera: the heart has in itself the *dunamis*, the power, of the demiurgic (*tên dunamin tên dêmiourgousan*) (647b5). While the previous model likened the generation of the viscera to the workings of a stream, this one asserts a demiurgic power behind their generation. We have, in this very brief section, two competing ways in which to understand the generation of certain parts—in mechanistic terms, such as the stream example, or by means of an appeal to demiurgic capacity.

The introduction of the heart allows us to turn to a discussion of blood and nutriment that will be combined with the elemental analysis brought in with the genetic framework (646a12 ff.).

4.4 Uniform Parts Within an Elemental Framework

The discussion of the heart is meant to overcome any tension between the sensitive parts and the instrumental parts.[12] Discussion of the heart, not unexpectedly, issues into a discussion of blood. Blood provides the means by which topics such as the hot and cold, fluid and solid, nutriment, and the passions are brought into the discussion. However, blood also provides more detail about the character of the uniform parts. The partness of parts is determined by the whole. This holds for uniform parts as well as non-uniform parts, like hands (Fig. 4.3). Just as a hand removed from a body ceases, in some sense, to be a hand, removing some of the uniform parts from the living organism will make them other—for example, the fluid will become firm (647b11, 649b30; Fig. 4.4). The whole is a cause of the nature and character of the parts, even those on the inside. Pointing out that the fluid uniform parts can become solid when removed from the living organism reminds us that these parts are close to the powers associated with the elements. Seeing the effect of blood removed from a body allows one to recognize the perhaps paradoxical complexity of the uniform. The solidification of blood points to the earthy material that makes up an apparently fluid part like blood. The hot is able to work on earthy elements of blood; as Aristotle says, the earthy elements can act like embers.

The uniform can play several roles when analyzing the parts of animals, but the present discussion disregards the role the uniform parts play in sensation. While the topic of sensation is initially introduced in *PA* II.1, a fuller treatment comes only with the new beginning of *PA* II.10. For now, we are exploring the uniform parts that are not sense organs. The uniform parts can be used as material for the non-uniform, the instrumental (647b22 ff.). Nourishment and therefore growth are accomplished

---

[12] Cf. Gill's (1989) treatment of paradoxical unity especially in terms of *tode ti* (esp. pp. 32–34).

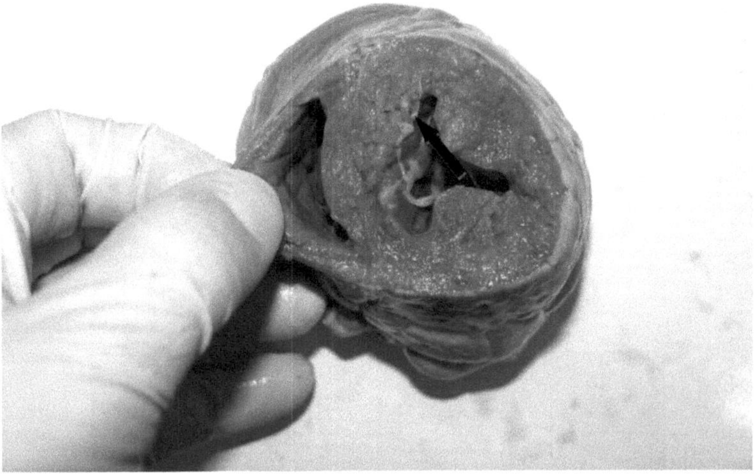

Fig. 4.3 In addition to the apparent external non-uniformity, Aristotle must have been struck, on the one hand, by the cavities present on the inside, and, on the other hand, by the uniformity of the muscle and tissues of the heart walls. The sheep heart pictured looks into the two ventricles with the apex cut away. The left ventricle is the more muscular one (*arrow*); Aristotle claims that the ventricle is dense in order to preserve the heat of the blood (*PA* 666a2). The external and internal characteristics of the heart allow Aristotle to put much importance on the heart. Aristotle claims that the hearts of the largest animals have three cavities (666b20–b35). Aristotle says that "it is better yet for there to be three cavities, so that there may be one, common origin; and the middle and odd-numbered one is an origin." I am inclined to think that he is taking the two ventricles together as "one cavity." From the outside, the ventricles look like one part surrounded by the two auricles—as seen in Fig. 4.2—while one can see the distinction between left and right ventricles in cross-section here. See Lennox (2001, p. 259) for other attempts to understand what Aristotle might have thought when claiming there are three cavities; see also Shoja et al. (2007)

by means of fluid uniform parts, that is, by means of the blood. Solid uniform masses are created in the form of fecal matter in the process of growth and nourishment. What is Aristotle suggesting about uniform parts when he includes excrement in that class? Excrement is the product of a process (e.g. 649a26); we can note in anticipation that blood is also the product of the process of concoction (650a25). One suspects that the uniform parts elicit the question of processes not immediately clear in examining the non-uniform. When examining something like a hand, one does not readily see animate process at work; one is inclined to see the function or work when one examines non-uniform or instrumental parts. Uniform parts seem to point to internal processes and motions while non-uniform parts draw one's attention to the external workings of the animal body. In the *Metaphysics*, Aristotle expands our notion of what it means to be made out of something. Substantial beings can be made out of material but also out of parts of the composite whole (*Meta*. 1035a18–22). The uniform/non-uniform discussion of the *PA* is, perhaps, meant to reflect the difference between matter and parts of a whole. While Aristotle

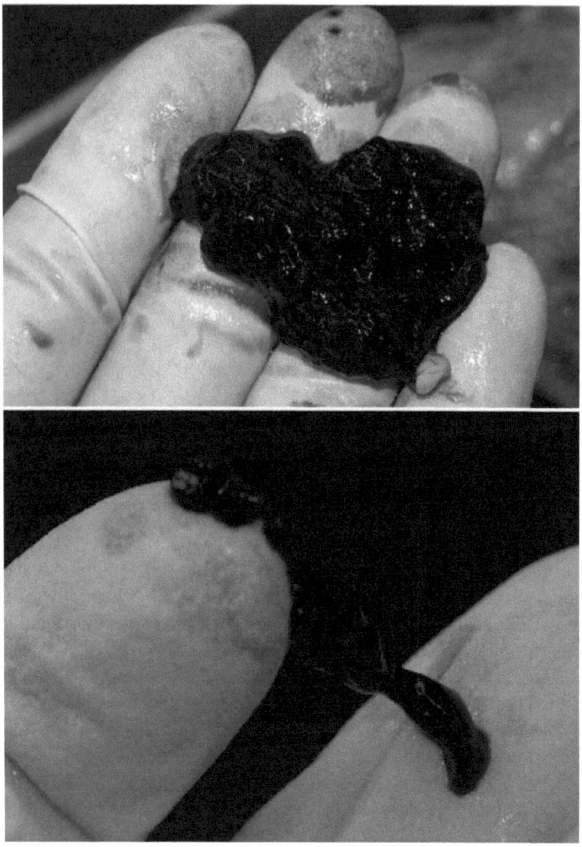

Fig. 4.4 Coagulated blood provided Aristotle with some evidence that this fluid element might contain an earthy element, which helps him explain, as we will see, the hotness of blood but also some elements considered psychological. The earthy element of blood only manifests itself when it ceases, in some important way, to be part of the organic whole. The heart has a two-ness associated with it in being both uniform and non-uniform. The complexity of blood is seen in its combination of the fluid and earthy, in addition to the hot. The hotness of the blood of the living seems to be replaced by its earthy character in death

indicates the incompleteness of an analysis that works its way up from the elements, in highlighting processes indicative of life we have one very important result of an investigation into the elemental and uniform.

Uniform parts can be distinguished by identifying differentiae (*diaphorai*). These differences can be understood to serve a purpose, they are for the sake of the better (647b29). Aristotle offers blood as an illustration. Blood is not only different in different kinds of animals, but can vary within a given organism. The differences in blood are manifested in what one might call differences in character. How the blood, or any uniform part for that matter, stands in relation to the elemental potencies has an effect on an organism's psychological makeup, so that thin and cold blood affect an organism's sensing and thinking (648a2). In this discussion, we see the argument moving towards an attempt to give a mechanistic account of certain phenomena usually associated with soul (see 651a12). This would be consistent with giving an account of anger in terms of the boiling of the blood (*De Anima* 403b1–2).

A mechanistic account of the characters of the organism must ultimately refer to the elements. This raises again the problem of trying to give an account of the animate in terms of the inanimate elements.[13] An understanding of the elements will also prove useful in attempting to understand processes like concoction, which results in the formation of blood. Before we get to such a discussion, an attempt has to be made to better understand the powers (*dunameis*) of the elements. Recasting the discussion of the elements in terms of power may be a way in which to address the problem of the relationship between the animate and inanimate. Perhaps it is possible to translate the inanimate elements into potencies that can be more easily, and more meaningfully, applied to the animate. Aristotle claims that the principles of the nature of the elements (*phusikon stoicheion*)—hot, cold, solid and fluid—are nearly (*schedon*) the causes controlling life and death, as well as states like sleep and waking, prime and age, disease and health (648b7). In an attempt to determine the "why," the cause of animals and their parts, we come close by appealing to the nature of the elements.[14] Of course, saying that we "come close" may be a way to indicate the failure or at least the limitations of such an endeavor; this provides some clue as the structure of *PA* II.

If it is correct to suggest that there is a very great difficulty in moving from the inanimate elements to life, one should emphasize that Aristotle does not simply reject the Pre-Socratic dependence on the physical elements and the elemental analysis; instead, he transforms the terms of the argument to suit his approach to animate things.[15] The elements are put to work, put into some kind of motion.

Plants and animals take up food or nourishment (*trophê*), transforming it into material used in the service of maintaining the whole. Food, in the form of the solid (*zeron*) and fluid (*hugron*) is subjected to the hot (*to thermon*) in the process of concoction or change (*metabolê*) (650a5). Fire—originally understood as an elemental, simple body—is transformed into a principle of heat that does a certain

[13]In thinking about Physics II.9, Cooper (1985) argues that "Aristotle contradicts the determinist views of the materialist philophers against whom his argument in the chapter is meant to be directed" (p. 161). He goes on to discuss how Aristotle accepts "Democritean" necessity and finds it figures into the account.

[14]Although he concludes that it is impossible to grasp living substance by means of the physio-chemical, Heidegger (1995) was impressed by the vigor and vitality of the pursuit of biological understanding. He saw in the re-invigorated biological sciences "a fundamental tendency to restore autonomy to 'life,' as the *specific manner of being pertaining to animal and plant*, and to secure this autonomy for it. This suggests that within the totality of what we call natural science, contemporary biology is attempting to defend itself against the tyranny of physics and chemistry (p. 188).

[15]Cooper (1985) suggests ways in which "material" or "Democritean" necessity is compatible with teological explanations.

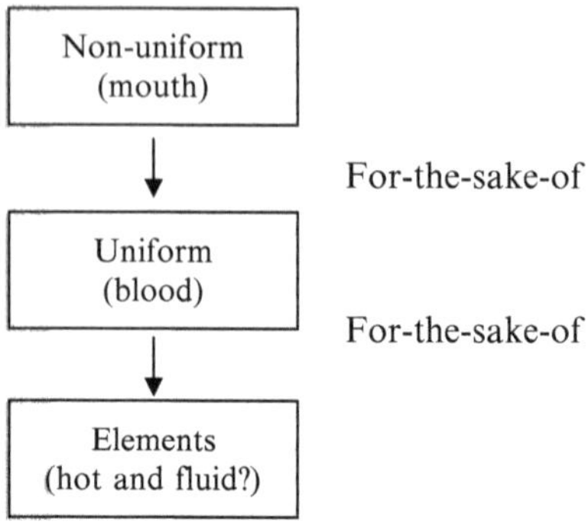

Fig. 4.5 The relationship between these three levels of composition as implied in the discussion of the process of concoction

work for the organism and accounts for certain aspects of generation. The processes implied in the terms usually translated as "concoction" (*pepsis* and *pesso*) are notions used in describing the ripening of fruit (*Meteorology* 380a11) or the cooking of food (*Mete.* 380b13). Concoction indicates change brought about by heat (*Mete.* 379b18, GA 775a17). The heat of concoction is an example of a notion used by Aristotle's predecessors in an elemental analysis translated into terms more appropriate for the animate. That the elements are, in a sense, brought into the organism is indicated by what Aristotle says in the context of drawing an analogy between plants and animals. Just as plants are rooted in the earth, getting much of their nutrients this way, the stomach in animals acts as if it were an earth inside the organism (650a24). In the course of the analysis, there is a sense in which the inanimate is drawn into the animate.

The introduction of processes like concoction, which result in the production of blood, has a bearing on the issue of the "for-the-sake-of" relationship between the uniform and the non-uniform. It was argued in at least two contexts that the uniform parts were for the sake of the non-uniform parts (646b7, 646b13; see Fig. 4.1). The discussion of the process of concoction, which results in a uniform part, makes it clear that certain actions of certain non-uniform parts are required for such a process. While non-uniform parts are not involved in concoction *per se*, they play a role in what might be called the preliminary stages, such as the mouth and the parts of the mouth that reduce solid food (650a9). But concoction is a process that yields blood and other uniform substances as an end (*telos*, e.g. 650a34). From this perspective, the mouth in mastication is for-the-sake of blood. This framework implies that non-uniform parts (such as the mouth) exist for the sake of the production or generation of the uniform (such as the blood). This is to reverse the for-the-sake-of relationship previously suggested (Fig. 4.5).

Inverting this relationship seems to show the complexity of the for-the-sake-of in living things. That the non-uniform seems to work for the sake of the uniform in the production of blood and nourishment seems clear. Just as the whole was absent in our working up from the elements, the elements are absent from this top-down analysis. In suggesting that the non-uniform is at work for the sake of the uniform is to say something like the spark of life—insofar as uniform blood is primarily hot—is preserved by non-uniform parts.

Blood, or its analog, is an end in the process of concoction but it is also a means by which nourishment is distributed through the animate whole, to the other parts of the body (650b1–11). In this latter capacity, blood can be good or bad, healthy or poor. The judgment that blood is good or bad is made when viewing blood as a means and the organism as a whole, since the process of nourishment is a process by which the whole is maintained. That the issue of nourishment points to the whole organism is also indicated by its role in the argument's new beginning of II.10.

4.5 Blood and the Passions

We have already seen how some of the elements, turned into potencies, apply to parts like blood. When blood is part of the organism it is fluid and is, at least in one respect, essentially hot (649b23). In addition to participating in these potencies, the fluid and the hot, the earthy or solid can be seen in blood insofar as blood can contain fibers (*is*). This term that Aristotle uses, *is*, comes to mean "sinews," "fibrous vessels in muscles" or "fibers" as it is used in this context. Homer uses it to denote "strength" or "force" of persons, such as Heracles, Telemachus or Odysseus (*Illiad* 23.720, *Odyssey* 2.409). Aristotle takes a poetic term and recasts it in physiological terms in this particular context. Aristotle had spoken metaphorically of the stomach as if it were an earth inside the organism (650a25); literally, organisms do have an earthy part in the form of fibers (650b18).[16] The notion that there are fibers in the blood is suggested by an examination of the cavities of the heart. Fiber-like bands connect the walls of the ventricle (Fig. 4.6). Both this observation and that of blood solidifying in coagulation (Fig. 4.4) suggests a combination of the wet and dry in blood.

Why should we wish to spend so much time and effort analyzing the constituents of blood? While blood offers an example, perhaps the best example, of the uniform parts that make up animals, it also highlights growth and nutrition. We would be remiss in our examination of the parts of animals if we did not thoroughly explore the uniform or at least the paradigmatic uniform part. Aristotle offers a much more compelling reason, however, for studying blood when he makes the claim that "it is well said that the disposition (*ethos*) and sensation of animals is caused by the nature (*phusis*) of blood" (651a12). Blood has a certain look, the look of a uniform part; this look is coupled with a recognition of the different potencies—fluid, hot, earthy—ingredient in blood. When stepping back and considering the organism as a whole, blood itself has a certain power that lies in its multiplicity of potencies. The look of blood is uniform, while it has multiple, heterogeneous potencies.

---

[16]On the significance of this observation, Ogle (1987, pp. 160–161, n. 1) notes that: "The coagulation of the blood is considered in two other passages (*HA* iii.6; iii.19). In all of these places Aristotle speaks distinctly of the coagulum as being formed of fibrous matter; thus anticipating the discovery usually said to have been made by Malpighi (cf. *M. Edwards, Lecons*, i.115). What Aristotle did not discover and what Malpighi did, was that by washing the coagulum the red colour could be discharged, and the fibres shown to be white."

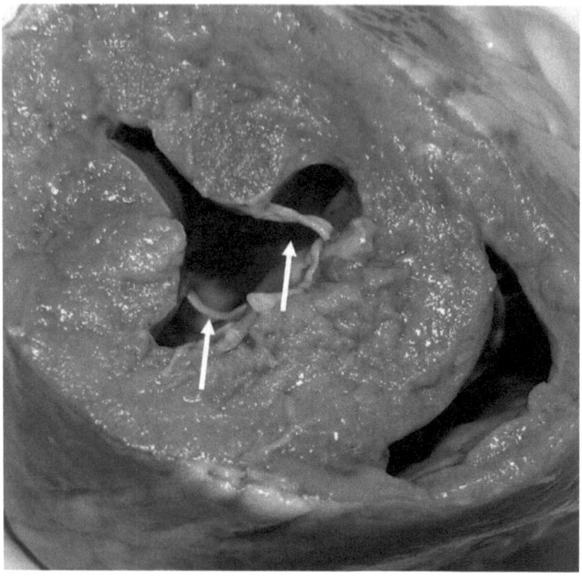

Fig. 4.6 The arrows indicate two of the fiber-like tethers present in the left ventricle. In *PA* III, Aristotle will explain the presence of the sinews in the ventricle of the heart as follows: "The heart also has many sinews, and this is reasonable. For the movements are from this part, and are accomplished through contracting and relaxing; so the heart needs such equipment and strength" (*PA* 666b13–16)

Certain passions seem to be rooted in blood type. For example, blood that is too watery produces cowardice because such blood has a connection with cold and the fear that accompanies cold (650b28, *Rhetoric* 1389b30). In contrast, blood that has an earthy character, with plenty of fibers, has a tendency to produce spirited or thumotic temperaments; the fibers act like embers, igniting fits of passions (651a1). The fluidity of blood is combined with heat and earthy fibers in this mechanical or material explanation of the passion of *thumos*. The account of the material cause of spiritedness and timorousness has the familiar Aristotelian structure of two extremes about a mean: the mean for blood seems to consist in being watery, while deviation towards being too watery results in cowardice and deviation towards the earthy results in a thumotic temperament.

The character of the blood affects not only the passions and temperament, but also the intellect (*dianoia*, 650b20). While the passions—timorousness and spiritedness (*thumos*) seem to be the examples to which Aristotle most often appeals—are most rooted in the temperature of the blood (the hot or the cold), the subtlety of an animal's intellectual capacity depends on the relative viscosity of the blood. The thinner the blood, the more mobile or kinetic an organism's sensitive capacity is (650b22). The duality of passion and intellect has a corresponding duality in the elements, since the hot and cold align with the passions while viscosity determines sensation and intellectual capacity. Such a picture is made more complicated when one recognizes that more viscous blood (i.e. blood with fibers) will produce a slowness of intellect, but also fits of passion as the fibers (that which is earthy) heat up like embers in the blood (651a3).

We have noted that the discussion of blood examines a so-called uniform part while also introducing the passions. In this discussion, we see more clearly Aristotle's transformation of the Pre- Socratics elements into potencies. The Pre-Socratic elements are supposed to be the material out of which everything is constructed; in the case of animals, Aristotle suggests, blood is the material (*hulê*) out of which the entire body (*pantos tou somatos*) is composed (651a14). However, blood is not simply a material: it is an end of the process of concoction, the process by which food is incorporated into the organism. In this role, blood is a means by which the whole organism is continuously nourished and thus maintained.[17] When Aristotle claims that blood is the material of the body, he is pointing to nourishment, to the fact that living things require nourishment in the maintenance of the whole. Saying that blood is material in this context goes beyond simply giving a material account; it points to generation and nourishment, which is the process of the continuous, material replacement of the organism.

The discussion of blood as the paradigmatic non-sensitive uniform part elicits several very important topics: it points to the elements in being hot, in having fluid and earthy characteristics; it points to genesis insofar as it is involved in nourishment and growth; and its effect on the organism brings up the topic of the passions of the soul. We will get to the topic of the soul again when the heat necessary for concoction is located. There seem to be two general principles regarding blood: (1) the whole body and all of its parts are composed out of and continuously renewed by blood (651a 14); (2) the peculiar nature of blood causes aspects of an animal's disposition and capacity for sensation (651a 21). Much of the argument of the first half of Book II can be understood as revolving around these two principles.

4.6 Contrarieties: The Hot and Cold, Blood and Brain

The transition from blood-like uniform parts (e.g. lard, suet, marrow) to a discussion of the brain involves a movement from the hot to the cold. There is an additional motivation for discussing the brain after marrow: Aristotle wants to dispel the notion that the brain is marrow (cf. *Timaeus* 75c–d), and thus in its nature blood-like. The

---

[17]The importance of blood and nourishment in this regard is perhaps close to Han Jonas' position when he suggests that "What, in its total effect, appears to be the maintaining of the given condition, is in fact achieved by way of a continuous moving beyond the given condition" (p. 197). This is in general an echo of Jonas' use of metabolism to establish certain elements of biological individuality: "... the mere phenomenon of *metabolism*, often considered to be no more than the elementary level that underlies, and supplies energy for, the quite different higher functions of life (such as perception, locomotion, and desire), in fact contains in its own primary constitution that groundwork as it were of all those functions, which may or may not evolve on its basis" (p. 196).

opinion that the brain is really marrow is supported by the observation that the spinal marrow is continuous with the brain. However, as Aristotle says, the brain and marrow are said to be of opposite natures (652a27). As a residue—a product of the process of concoction—marrow is hot and gets its heat from this process. In contrast, the brain is the coldest of all the parts in the body (652a28). The nature of the brain lies in the cold.

But we might ask ourselves, shouldn't all the parts of the body participate in the hot insofar as they are constructed out of blood? In the discussion of spiritedness and cowardice, we saw how the earthy, the fluid and the hot, the latter being primary, all combined in the examination. In other words, three of the elements, or potencies, were employed. With the brain, we are introduced to a fourth potency. The question becomes, is there a place for the cold in a body composed entirely, at some level, of hot blood? The brain is not a residue because if it were, it would be hot due to the process of concoction. Instead, "it is well said that the brain is peculiar in nature" (652b). Like the residues, however, the brain produces no sensation when touched (652b7).

The brain is not a residue and thus not part of the hot, providing balance to the whole organism. Blood and the other residues infuse the body with heat. In order that the organism, as a whole, may achieve moderation and the mean (*metriou kai tou mesou*), the cold brain—water and earth in common (652b22, 653a22)—provides a counterbalance to the heat of blood. For this reason, Aristotle is able to claim "the brain is present in order to preserve the animal as a whole in its nature" (652b7). In *PA* III, the cooling function will be located in respiration and the lungs.

We have confirmed that the heat ingredient in blood and the other residues is the result of the process of concoction. In the context of introducing the brain as the source of coldness, Aristotle identifies the soul as the source of the heat seen in concoction (652b8–14). Apparently, recognizing heat or the hot as a principle of soul has led some thinkers to claim that the soul is fire (652b8, *De Anima* 403b31). Saying that the soul is fire is similar to the attempt to identify the sense organs with one of the elements (647a12) and is generally consistent with the attempt to reduce everything to elements. Claiming that the soul is fire implies that it is homogenous like the sensitives (647a12). However, according to Aristotle, the soul is like a craftsman or technician that uses fire, as one performing an art would employ an auger (652b15). Such an image makes one pose the question about the part-ness of the soul and its work or function. Aristotle suggests that heat is present due to the work (*ergon*) of the soul including nourishment and motion. Such work is best initiated by the potent heat (652b13). The cold nature of the brain, sharing in common with the watery and earthy, is set up as a contrary to the fiery soul. Parts with contrary natures, parts that are in tension, allow for the well-being of the whole.

However, the brain is not totally isolated from the heat of the heart's blood. Aristotle explains that there are a number of small blood vessels embedded in the membrane around the brain (652b30); while the brain is protected by a barrier of sorts against damage from the heat of the blood, the blood in the membrane also allows for a moderate amount of warming. The brain has to be protected from the hot blood because there is a certain movement of heat from the area around the heart up toward the brain, which helps to explain the origin of fluxes (652b34). The abundance of heat of human beings is also used in the account of our upright posture (653a31–33).[18] The heat associated with the concoction of nourishment flows up toward the head and if in this process the brain gets too cold or deviates from the rightly proportioned blend, a flux will result. The word that is used for this medical condition, *to rheuma*, is the same word used to describe "that which flows," or "a stream" (cf. *Timaeus* 44b, 45c). The process of heat flowing upward in the organism, Aristotle says, resembles the genesis of rain (653a4); vapor is carried up from the earth by the heat and falls back in the form of rain once it reaches the coldness of higher altitudes. Aristotle indicates that this explanation might belong to natural philosophy (*phusikês philosophia*) (653a10).[19] An analogy is drawn between an organic process and a mechanical one which has the appearance of an account Aristotle's predecessors might offer. In the context of a discussion of the brain and processes associated with it, Aristotle debunks the idea that it is the gods who are responsible for rain. This is consistent with the overall character of the first half of Book II, in contrast to the appeal to the divine that initiates the new beginning in *PA* II.10.

We have noted that the cold brain is meant to counterbalance the hot blood and heart, in order to reach some kind of mean or moderation. This state of moderation is said to be a cause or to possess substantial being (652b20). Is such a state meant to be thought of as something like homeostasis? Should the study of the mean be included in an attempt to acquire knowledge of cause? In the organismal body, some balance is continuously maintained between the hot and the cold. In addition, for the functioning of the brain, a balance between the fluid and solid must be maintained (653b4). An organism is made up of parts that may be at odds but are dynamically balanced. A certain tension in the parts and in the nature of those parts seems to be necessary. If this tension is overcome by one part becoming dominant, such as lard and suet taking over the uniform sensitive parts (651b), then the whole will be undermined and the organism will perish.

4.7 From Inside to the Outside to the Inside: The Primacy of Flesh and Touch

The argument first considers fluid uniform parts in all their manifestations (653b9). However, the non-fluid parts must still be considered. The fluid uniform parts are primarily tissues and things inside the body, those things not evident to the senses. The move from the wet to the non-fluid is, in a way, one from the inside to the outside.

Flesh (*sarkos*) is the part—it does, at first, seem counter-intuitive to call it a part—that covers the whole, it defines the outline of an organism's form. This is

[18] Lennox (2001) describes this as the "thermomechanical explanation" of man's upright posture (p. 211).

[19] We can note that the parallel drawn between animal physiology and the mechanics of rain is very similar to the way in which the Aristophanic Socrates describes rain and those things associated with rain to Strepsiades (*Clouds* 367 ff.).

why Aristotle can say that flesh is both a principle and a body itself (653b23). Flesh provides a kind of limit to the unlimited fluid parts like blood, fat and marrow. In a crude sense, flesh forms a bag or container by which the internal parts are limited.

In the beginning of *PA* II, we were treated with two different ways to differentiate the parts of animals: the division into three classes based on genesis (646a12) and the division of parts based on the instrumental and sensitive (647a3). In flesh, we have a part that overcomes the split between these two methods of differentiation. Flesh is used as an example of the uniform in the genetic account (646a3), as well as being treated as an organ of sense. Flesh is the part through which touch is mediated, which Aristotle identifies as the primary sense. And as we have already noted, flesh—unlike ears and eyes—is the organ of touch as well as the medium (653b25). Flesh must come in contact with that which is to be sensed; there is no distance in touch as in the other sensitives. Nature made touch in this way, Aristotle claims, because of necessity (653b29).

Our discussion of the parts that are on the outside is brief in this context. With the discussion of flesh more or less complete, we turn once again to the inside, to the system of bones. The bones, which are in their nature hard substances, exist for the sake of the preservation of the soft parts (653b33). Just as organisms must combine the hot and cold, so too must the hard and soft be combined in the composition of their parts. The role assigned to the bones is perhaps best illustrated in the case of things that have exoskeletons, which also explains how the discussion can turn from flesh to bone. In most animals the soft parts or flesh cover a bony skeleton, but some organisms, such as the various shell-skinned (*ostrakoderma*) creatures (654a2) have the reverse relation, namely, a hard substance encasing fleshy parts. The shell not only provides protection for the fleshy innards, but as Aristotle conjectures, it also insulates and protects a faintly burning heat (654a8). While the flesh functions as a bag or limit as well as an organ of sense for birds and mammals, the outer covering provides protection to the fleshy parts and to the heat generated for those organisms with shells. In most of the organisms that we come in contact with on a daily basis, such as dogs, cats, birds, and other humans, the bones are internal.

The case of bones, and even blood vessels, provides a very illuminating example of the relationship of parts to wholes. "A bone on its own," Aristotle observes, "is nothing; rather, it is a part either as a part of something continuous (*sunechês*) or through contact and binding, in order that nature may use it both as one and continuous and, for bending, as two divided" (654a34–b3, Fig. 4.7). Individual bones make up a larger system of the whole of bones. In one respect, bones are just one part of the organism as a whole; in another respect, that part is an articulated whole, the skeleton. The radius and ulna are arm bones that are, in a way, parts of two wholes, the skeleton and the organism. There exists a whole within the whole.[20]

[20]On this general point, Furth (1987) remarks that biological individuals "display to a marked degree a hierarchical structure of *levels* of organization, in which what is a 'whole' at one level is a 'part' at the next, and in which the part-whole relationship itself assumes a variety of forms beyond that of ingredients in a mixture, or aggregation into a bulk or a 'heap'" (pp. 27–28).

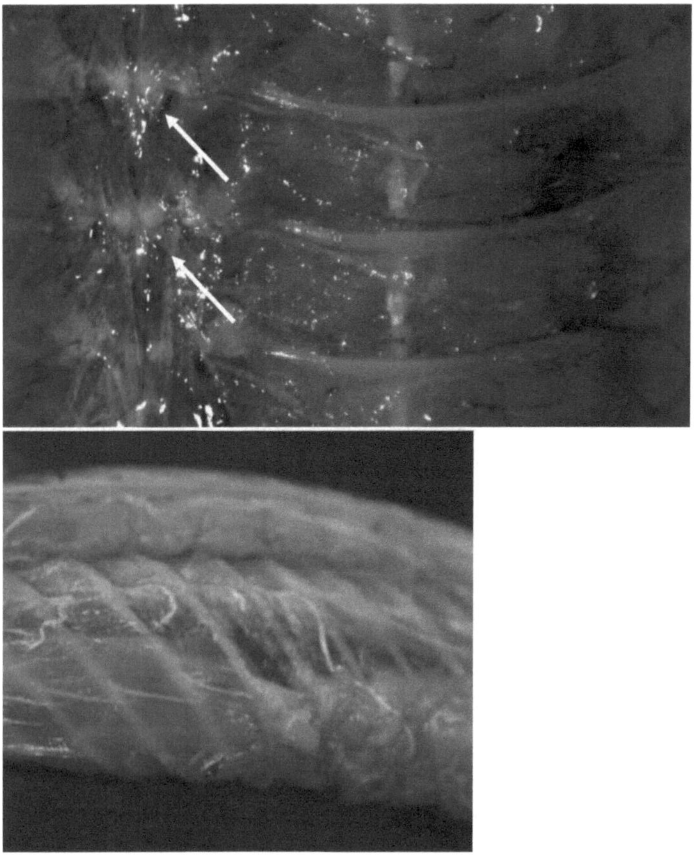

Fig. 4.7 The vertebrae (*arrows*) of the backbone and ribs of a black snake from inside the body cavity (*top*) and from outside offer a good example of Aristotle's claim that the backbone "though one on account of its continuity, is made of many parts by the division of the vertebrae" (*PA* 654b15). The ribs are each connected to the backbone while at the same time being connected to one another by means of the "contact and binding" of muscles and sinews (*PA* 654a35). The ribs are parts in that they are connected to the continuous whole of the backbone; but they are also parts by being connected to one another through the binding and plaiting of soft tissue

Also, just as the system of blood vessels has as its source the heart, the system of bones is a connected whole or holding together that originates with the backbone (654b13).

Aristotle makes a very interesting connection between, as he says, the "nature of bones" and the "nature of blood vessels." The likeness of blood vessels and bones is not immediately obvious. Both blood vessels and bones have an origin (*archê*); as the blood vessels are to the heart so to are the bones to the backbone. These systems or suites of parts are understood as alike on the basis of analogy. While we originally thought that analogy was to be helpful in talking about feathers and scales, here we see analogy applied to parts of the same whole. The heart is the first organ to move while the backbone and bones allow the vertebrate, organismal whole to move in the world. Further, the backbone, Aristotle notes, is "one on account of its continuity" while it "is made of many parts by the division of the vertebrae" (654b15–16). This character of the backbone is similar to the observation that the heart shares in both the uniform and non-uniform.

On the one hand, the backbone is a holding together, it is continuous; on the other hand, it must be divided, because of the necessities associated with animal motion (654b15), into a many-parted-thing (*polumeres*). Motion requires that one have a continuous whole or a one that can be cut up or articulated into semi-discrete units. However, a bone that is truly isolated, disarticulated, is not able to do its work (*ergon*) and in fact might be a source of harm (654b5). Individual bones are not isolated but can be articulated at the joints; the presence of joints connecting individual bones allows for a variety of movements (654b20). The backbone is a one that can be divided into many, allowing for a plurality of motions.

Not all bones move. Just as a hard, bony substance provides protection in shell-skinned creatures, those bones that do not move, e.g. the ribs, can serve as a means of protection for the internal viscera (655a). There are some soft parts, however, that need to expand and bones would hinder such an expansion. The belly, for instance, needs to expand both in order to accommodate food, and, perhaps more importantly, to accommodate the fetus in viviparous animals. Bringing up different modes of reproduction allows Aristotle to make a digression on the nature of bones in relation to the different lifestyles of different kinds of animals (655a5–655b2). This digression is followed by another on parts that are like bone in being hard (655b2–655b28).

With the observation that bones might hinder the growth of the fetus in viviparous animals, Aristotle relates aspects of osteology and morphology to life history. He observes that live-bearing creatures are usually larger in body size and require stronger, bigger and harder bones (655a12).[21] Despite living in water, dolphins, and presumably other cetaceans, have bones like their live-bearing kin, as opposed to what Aristotle calls "fish-spine" or the cartilage of the cartilaginous fishes (655a23). With regard to the nature of their bones, dolphins are united with other viviparous organisms.

Because of the nature of their movements, selachians (sharks and rays) have cartilaginous skeletons which are somewhat pliable; this plasticity, Aristotle suggests, facilitates the undulating quality of their movement (655a24). Organisms that have true bone forming their skeletons can also have cartilaginous parts as in the case of ears and noses (655a31). Cartilage and bone are the same in nature,

[21]D'Arcy Thompson (1942) supplies evidence for bones being correlated with body size when he notes that bones make up 8 percent of the body of a mouse, 14 percent of goose or dog, and 18 percent of the body of a man (p. 28).

but are differentiated by "the more and less" (655a33). Whereas animals with cartilaginous "bones", e.g. sharks and rays, differ from animals with "true bone" by more than the "more and the less," bone and cartilage differ only by the more and the less.

Aristotle makes a distinction between the bones of carnivores and those of herbivores within the viviparous (655a13). While both are constructed out of true bone, the bones of a carnivore are necessarily harder than the bones of an herbivore. This can be explained by a difference in the way each gets food. In this digression on certain aspects of bone, Aristotle attempts to understand an organism's osteological and morphological characters with respect to aspects of body size, habitat, mode of locomotion, and other life history characteristics. In asking these questions and making these observations, Aristotle offers a causal account based on the work or life history of the organism: the carnivores have harder bones *because* of the way in which they get their food. This seems to be a case in which *historia* provides a causal explanation which is significant in thinking about the relationship between the *HA* and the *PA* with respect to determining cause.

4.8 New Beginning: The Polymorphic or Polyeidetic

The digression into bones and lifestyle prepares, in a way, for the new beginning that the argument now takes. Our new beginning is to start from the first things (655b28). The necessity of food and nourishment in the growing and maintenance of an organism appears to be among the first things, since the new beginning is initiated by an examination of the necessity of nourishment (655b29). This is not to suggest that the topic of nourishment was not ingredient in the preceding discussion; but that discussion was always grounded in the analysis of blood, which involved the elements or potencies—hot, cold, solid, fluid.

With regard to nourishment and growth, Aristotle tells us that among the most complete creatures (*zoiois tois teleiois*) there are two parts that are most necessary (655b30)—that by which food enters, i.e. the mouth, and that by which residues exit, i.e. the anus. In this formulation we have a mixture of the final (*teleos*) and the necessary: the final applies to the organism as a whole (*zoios tois teleiois*), while the necessary applies to the parts of that organism (*anagnkaiotata moria*) in the face of the need for nourishment and growth. If one were to take the mean between these two most necessary parts—mouth and anus—one would discover the principle of animal life (*hê archê estin hê tês zoes*) (655b39). The doctrine of the mean applied to two holes yields the seat of life. Though this may be meant as a joke, these two most necessary parts highlight the process of nourishment and growth, the struggle for existence and the maintenance of the living organized whole. As I am inclined to argue, the beginning of *PA* II does not attempt to account for the whole organism insofar as the analysis is grounded in the inanimate elements and their place in *sunthesis*, or *genesis* as *sunthesis*. With the new beginning and its emphasis on the completed (whole) animal *(zoios tois teleiois)*, including its nourishment and maintenance, we are able to view the phenomenon of life from a different perspective.[22]

While the distinction between perceptual and instrumental parts had been introduced early in *PA* II (647a3), it had not been put to use in the analysis in any significant way. Something in the change in the argument now allows for, or makes necessary, a discussion of the organs of sense. In addition to the struggle involved in living, some organisms possess the power of sensation which comes with a polymorphism of looks (*polumorphoteron idean*) (656a4). Plants are not polyeidetic (656a), in part because of a lack of sensation, and in part because they have few actions to perform (656a3).[23] The move from plant life to sensitive life is not accidentally accompanied by polyeideticism. The diversity of form or polymorphism is increased in those kinds which not only sense but have the capacity to live well (*eu zoên*) (656a7). Human beings, in part because they have the capacity to live well, show up in the greatest variety of forms. But polymorphism, according to Aristotle, is a sign of divinity: "humans are the only animals known to us to have something of the divine, or at least if there are other animals, he is the most divine" (656a8; see also *PA* 686a25).[24] The polymorphism associated with some aspect of divinity is the real motivation for our "new beginning." This echoes an important theme first articulated in *PA* I; the animal body, Aristotle suggests, is for the sake of a many-parted action (645b17) and is, in a way, also for the sake of the soul (645b19; see also *De Anima* 415b17–21). Emphasis in this context is on the polymorphism; the polymorphism of humans is coupled with a potential for many-parted actions. Polymorphism is evident in both look and power. One wonders if the beginning of *PA* II grounded in the elements would then be contrasted with the beginning in the polymorphic divine. I want to suggest that we are meant to better understand the animate by means of a contrast with these two poles, with the inanimate elements of the first half of *PA* II and the divine as the investigation shifts to the second half. In this way, the animate is viewed against the sub- and supra-animate.

When speaking of the elements, Aristotle had remarked on the necessity of having dissimilar potencies (*dunameis*); various potencies are required because of

[22]Gotthelf (1987) understands this passage as assuming some demarcation between the "lowest of the low" and "higher" animals, between plant-like animals and what we might distinguish as true animals (see p. 182, n. 43). While I am sympathetic to Gotthelf's interpretation especially in attempting to account for what is to come in the argument, I am more inclined to view the statement regarding the "perfect animals" as pointing to the organism as a whole: this seems to me to explain better the movement of the argument of Book II, from the glaring absence of the whole in the first half to its initiating a "new beginning" in II.10.

[23]The polyeideticism of the animal kingdom might be consistent with Jonas' (1968) concept of centralization: "The stationary plant could no more profit from *centralization* and individuality than the moving animal could be without them. We see accordingly that centralization is not the same as animate *unity* of the complex whole, nor always accompanying such unity, but is a new fact in the evolution of the metazoic (multi-cellular) organisms, confined to animal life and coincident with the evolution of sentience and motility" (p. 198).

[24]cf. Plato's *Sophist* 271e7, *Statesman* 256b2–4.

the polymorphic nature of the actions and movements of whole animals (646b15). The movement of the argument of Book II seems to be one from the varied and diverse potencies of the elements to the varied and polymorphic nature of creatures with the power of sensation, especially human beings in their capacity to live the good life. Polymorphism will be a way in which Aristotle can introduce certain multiplicities, first in the sense organs or body in general, then in function at the level of the parts and at the level of the organism as a whole. This latter task will be accomplished by examining functions that are common, those according to *genos* and more particular ones; while not exploring it here, it is worth noting that this thread of the analysis continues into *PA* III and IV.

Our new beginning in *PA* II.10 is meant to refocus our examination on what comes first (655b31). In addition to the divine polymorphism of humans, one reason we should begin again with humans is the familiarity of our own external parts (656a7). We are more familiar with the human shape. It is from a combination of internal—humans having something divine in them—and external—human form being most familiar—that we should begin again with humans.

One of the striking features of the human being is, according to Aristotle, the fact that we are the only animals to stand correctly, to have our natural parts (*phusei moria*) according to nature (*kata phusin*) (656a11). In humans, the upper part is situated towards the whole (*pros to tous holou*) (656a12).[25] In this context, the whole to which Aristotle refers is not the animal but the cosmos. We can start our new beginning with humans because they have something divine in them and because their stature allows for an examination of the whole, the cosmos. In a certain respect, the unique stature of human beings appears to be, as Aristotle suggests, the byproduct of the mechanical workings of the elements. Heat rises, and the fact that humans have so much heat and blood in the region around the heart induces growth upwards along a center line (653a30). What, then, is the cause of our stature? Is it for the sake of looking at the whole cosmos or a result of our abundant heat or both at once?

Whether or not Aristotle refers specifically to the head and the parts situated in the head as the "natural parts" that stand in a special relationship to the whole, the head is more peculiar than any other part (656a26). The peculiarity of the head, with its relative lack of flesh, induces some to believe that the brain is responsible for sensation, while Aristotle suggests that the seat and source of sensation is the heart (656a27, cf. *De Sensu* 438b25). Aristotle argues against the lack of flesh around the head being for the sake of the facilitation of sensation; instead, the relative lack of flesh is explained by the brain's activity as a cool counterbalance to the hot blood and heart (652b6, 656a20).

[25]Erwin Straus (1966) puts a great deal of emphasis on the upright posture of humans: "Obviously, upright posture is not confined to the technical problems of locomotion. It contains a psychological element. It is pregnant with a meaning not exhausted by the physiological tasks of meeting the forces of gravity and maintaining equilibrium" (p. 137).

If Aristotle really wishes to identify the seat and source of sensation as being somewhere other than the head, he has to explain why the organs of sight, hearing, smell and taste are all situated as parts of the head. The cause of the placement of at least three of the organs of sense about the head rests on the nature (*phusin*) of these sensitives (656a34). Sensation works best where the blood has less of an effect. The heat and the movement associated with the hot impedes the being-at-work (*energeia*) of the sensitives (656b5). Unlike touch, which deals in oppositions (such as the hot and cold, fluid and solid) (647a18), hearing and especially sight make finer distinctions. The sense organ of sight is water in its nature (656b) and the sense organ of hearing is air (656b16). The *energeia* of these sensitives, which are most precise, is the cause of their being situated around the brain.

In their attempt to reduce everything to elements, Aristotle's predecessors identified each of the sensitives with an element, asserting one is fire, one is air, etc. (647a12). Far from rejecting this Pre-Socratic effort out of hand, Aristotle identifies the sensitives with elements, the organ of sight with water (656b), and the organ of hearing with air (656b16). In contrast to the Pre-Socratics, though, Aristotle's appeal to the elements in accounting for the majority of the sensitives being in the head takes its bearing from *energeia* as a cause; it is something more than merely stating a material cause.

While the sense organs were introduced among the uniform parts (647a6), they are not addressed in an adequate way until the latter half of *PA* II. This seems to be a result of replacing the framework that focused on parts as elements, uniform, and non-uniform (641a12) with one focused on the division between instrumental and sensitive (647a3). The latter way is apparently more illuminating for the study of the animal as a whole, since the perceptual parts, such as the eyes, are parts, which more than other parts show the work of the animal as a functioning being in the environment.

4.9 Doubleness

When contrasting the instrumental parts with the sensitives, Aristotle approves of the suggestion that sensation takes place in simple (*haplos*) parts (e.g. 647a14). When the topic of sensation returns in the context of our new beginning, he claims the sensitives are double (*diplos*) (656b34). The presentation of the senses as simple apparently needs to be re-examined. The sense organs (the ears, the eyes and even the nose) come in pairs. These perceptual parts are literally two-parted (*dimeres*) (657a30). As we shall see, the many-ness of the organs of sense extends beyond the fact that they come in pairs.

Of the senses, touch is the only one, if one accepts forked tongues, that appears to be undivided. Still, there is a doubleness to touch that, while not of the same double character as of paired eyes and ears, seems to give it a privileged status. That doubleness is the duality of outside and inside (656b36). In the earlier discussion of flesh, Aristotle identifies it as being a kind of body in itself, but also the part by which touch operates (653b19); in the context of discussing the doubleness of the organs, this role of flesh is not mentioned; instead, Aristotle says that the primary (*proton*) sensitive in touch is something internal (656b36).

The nose also offers an example of doubleness very different from eyes and ears. The nose is divided in that it has two passages, two nostrils, but the nose is not separated into two, like the ears (657a4). A more significant duality of the nose lies in a duality of function: the nose is the means by which we smell, but it is also a part by which we breathe (657a10). As we shall see, the nose of the elephant will illustrate an even greater multiplicity of functions (658b33). One wonders to what extent the doubleness of function in the nose is applicable to the other sensitives. For example, the ears and hearing alert an organism to noises in its environment; but in humans in particular, the ears function in language activities. Much of the second half of *PA* II involves the discussion of organs and parts that come to have multiple functions. The movement of the argument progresses from a study of the fact that the sense organs are often double, to an examination of the doubleness of function.

The function of particular parts and especially sensitives are discussed in terms of an organism's peculiar life history. Aristotle says that the sense organs of animals are beautifully arranged in the face of their idiosyncratic natures (*idian phusin*) (657a11). The parts are not simply beautifully arranged in and of themselves, but in relationship to a particular nature, a particular way of life or being-at-work. So for example, the ears of tetrapods are situated in such a way that they are useful for, adapted for, their movements (657a13).[26] Aristotle suggests elsewhere that variations (*diaphorai*) in parts either have some reference to the work (*erga*) or being of the animal or are better or handy (648a15). This statement made with regard to uniform parts is now applied to Aristotle's discussion of the parts associated with sensation.

Sense organs are handy in relation to the work of the organism, but they are under certain material constraints. For example, birds do not have ears *per se* even though they participate in hearing; instead, they have auditory passages as a result of the hardness of their skin (657a18). While the ears of tetrapods are offered as an example of the useful, the auditory passages of birds are offered as an example of the constraints of material necessity. In addition, the four-footed animals that lay eggs also do not have "recognizable" ears as a result of material necessity. It is in this context that Aristotle makes a very curious note regarding the seal (Fig. 4.8): "among the live-bearing animals even the seal has, no ears, but auditory channels, because it is a deformed four-footed animal" (*PA* 657a22–24). Seals are said to lack ears because they are deformed quadrupeds. I would have expected an account about the lack of ears being related to the seal's way of life. At the very least, based on the material account regarding the egg-laying tetrapods, I would have expected Aristotle to say that the material for ears was used elsewhere. Saying that seals are

[26]For a discussion of adaptation in Aristotle's biological works see Balme 1987a, b.

Fig. 4.8 Egg-laying tetrapods such as tortoises do not have ears, but canals. This is on account of the hard, scaly material that makes up their outer covering. Likewise, seals have canals that function as ears. Notice the well-developed flippers (*bottom*). Such an observation might be folded into an account that puts emphasis on the material required for the "deformed" flippers at the expense of ears (cf. Lennox (2001) p. 228)

Fig. 4.9 The eyes of a mantis shrimp (*Squilla mantis*) offer a good example of Aristotle's discussion of the hardness and mobility of the organ in hard-skinned animals. Thompson (1947) argues that the squilla are equivalent, or nearly so, to the *karis*, small crustaceans other than crabs and lobsters (p. 103)

deformed, might point to the fact that their flippers are developed from cartilaginous material to such a degree that there was not enough for the shaping of proper ears. Speculating along these lines would help make sense of Aristotle's remarks about the material constraints in tortoises and his notes on seals in the very same context.

In the course of discussing the eyes, Aristotle gives an account of eyelashes and eyebrows, which are likened to architectural structures; we are told that they are for the sake of protection, that their purpose (*heneka*) is to keep things out (658b). But we also learn that the eyelashes and eyebrows exist because of the workings of material necessity (658b20). In rapid succession, Aristotle gives an account of hair based on material necessity and one based on some teleological concerns (658b3–b7).

It is easy to take for granted that eyelids and eyebrows necessary accompany eyes, but that is not the case with hard-skinned organisms such as insects and crustacea. In order to compensate for the lack of eyelids, such organisms have hard eyes, which might be understood to combine the eyelid and eyeball (Fig. 4.9).

Aristotle weaves together the useful and the necessary throughout this inquiry into the parts of animals, which is to follow the way outlined in *PA* 1 (642a2). At the same time, he analyzes the way in which an organ with a more common function across kinds can be pressed into more particular tasks depending on life history (645b21), as is illustrated by the nose of an elephant.

4.10 Variations on a Theme: The Handiness of the Elephant Nose

In the preceding section, we examined the way in which elements of the argument bring together the necessary and final causes, a way of proceeding suggested in *PA* 1. This is followed by an examination of the way in which attributes can be common, according to *genos* or according to *eidos*, which was another way the argument was to precede, according to the methodological approach sketched at the end of *PA* I (645b25). The nose, Aristotle now proposes, varies little among viviparous tetrapods (658b27). However, one would suspect that any variations in this part were ingredient in the animal's substantial being (*ousia*) or are beneficial or handy (648a15). It turns out that the variation of the elephant from the general conception of nose is very handy indeed. The elephant's nose is idiosyncratic in being large and extraordinarily potent (*dunamin*) (658b33). A term used to denote the powers of the elements is now applied to particular parts of an organism; this change seems to be consistent with the movement of the argument, as I have outlined it. The nose usually has a double function, as an organ of smell and respiration (657a4); these are functions that are common across a wide variety of kinds. In the elephant, another function is observed in its use as a hand (659a2). It can function as a hand would function because it is prehensile, a quality that is dependent on its material make-up (659a16). What is normally thought of as a sense-organ is shown to be an instrumental part in a sense, thus suggesting the dichotomy between these two, the sensitive and instrumental, has been qualified and refined. The normally passive quality of the sensitives is replaced in the elephant's nose by an instrument that can be utilized to tear up trees. Because of its size and its prehensile nature, this one part has an incredible variety of functions that transcend the instrumental/sensitive, uniform/non-uniform dichotomies. In addition, Aristotle's analysis has demonstrated that there is not a one-to-one relationship between part and function.[27]

An elephant's nose is used to convey both solid and fluid (*zeran kai hugron*) nourishment to the mouth; this description reminds us of the way in which the elements were used in the discussion of nourishment and blood in the first part of *PA* II. But fluid and solid are also descriptions of what we would refer to as the elephant's habitat; he is a land animal, but also is swampy (*helôdes*); an elephant is furnished with what is useful, which is the same as what is necessary, for life in water and on land (659a7). Despite being too large to move easily between water and land (659a5)—a fact which in and of itself may argue against some large-scale teleological design—the elephant is suited to live the peculiar kind of life that it does. Through art, humans can construct a machine (*organon*) by which they can

[27] See Tipton (2001) for a discussion of the relevance of Aristotle's recognition of the division and combination of labor of animate parts to certain contemporary issues.

do what elephants do naturally, maintain a connection to the air we breathe while submerged under water (659a9). Humans can make a machine which mimics the function of an organ of the elephant. In fact, the nostril, normally an organ of smell and respiration, becomes an instrument for the elephant that allows for varied functions. The instrumental and sensitive are combined in the trunk, which yields an array of functionality.

That the emphasis over the course of *PA* II shifts from the diverse powers (*dunameis*) of the elemental to the diverse and multiple functions of the instrumental parts, and the organisms which possess such parts, is clear; but that shift is not as abrupt as one might expect with the announcement of a new beginning. The movement of the argument toward a recognition that parts can serve multiple and idiosyncratic functions is anticipated in several ways before the explicit turn. A discussion of the bony, hard covering found in some organisms provides one example that came before the new beginning: oysters and crustaceans have shells that not only protect the soft parts, but function additionally in the preservation of heat (654a). Another example of this multiplicity of function involves the role teeth play in mastication as well as defense and self-preservation (655b9). In fact, function can delineate a part without the part actually being present in any ordinarily conceived of way. For example, birds have only what can be called "nostrils" based on function (659b); there is no readily identifiable nose or nostrils on birds, yet something functions in order that birds may smell or do what noses normally do (Fig. 4.10). Function, in contrast to look or morphology, determines whether the part is present.

The discussion of the elephant's trunk is embedded within an account of the nose, which in turn is part of a discussion of taste. Just as we discussed things like eyelashes and eyebrows in the account of eyes (e.g. 658b14) we turn our attention to an examination of the work associated with those parts around the sensitives as well as those parts directly responsible for sensation. From smell, we turn our attention to taste, which prompts a discussion of the parts associated with the mouth. As we learn, the mouth, especially the human mouth, will yield a number of distinct parts; and like the elephant's trunk, the mouth will yield distinct and numerous functions, the most important of which includes the production of human speech. Teeth can provide a means by which an organism defends itself (655b9) although the teeth themselves need protection. "In other animals (besides humans)," Aristotle says, "the lips are intended to preserve and protect the teeth" (659b28). But while human lips serve to protect the teeth, they perform a more important function in the production of speech, which is according to some good. (659b33). This doubleness with regard to the functioning of human lips is also seen in the tongue, which plays a role in taste, but also in the production of speech (660a).

There are several distinctions illuminated by the discussion of lips, tongue, and teeth in the context of human speech. In each case, a part is shown to have multiple functions. The idea of a one-to-one mapping of part and function is broken down at several levels. If human speech is a work or function of human beings, then it appears as if no one part is responsible for this *ergon*. Not only do parts (for example, the teeth) have multiple functions, but also functions (for example, speech) require

Fig. 4.10 Neither the holes on the buzzard (*Buteo buteo*, *top*) nor the jackdaw beak (*bottom*) look like nostrils, yet they are functional as recognized by Aristotle. Their proper function, and not their look, determines their status as nostrils

the working of multiple parts. The multiplicity of function evident in the animal world corrects something about the craft analogy; an ax has one function and to use it for another, say, shave, would be a violation.

4.11 The Interweaving of the Material and Teleological

What is illuminating about the argument of the first half of *PA* II is its emphasis on the material make-up of the animate; what is absent is some account of the whole, some understanding of how the various uniform and non-uniform parts for a unified whole. As the argument moves from *PA* II.1–9 into Chaps. 10–17, the instrumental/sensitive distinctions get divorced from the uniform/non-uniform distinction; in other words, our primary concern is no longer whether a sensitive or perceptual part is uniform, with its material constituents. It is not enough to dwell exclusively on the material, nor should such an inquiry be bypassed. For example, the discussion of the elemental qualities of blood directs one's attention to indispensable processes such as growth and nourishment. Additionally, what looks to be unvarying homogenous stuff like blood, will on further analysis yield variations that help understand the diversity of animate life. That the dialectical move from the material to the animate whole is necessary for our understanding is evidenced by the structure of the argument of *PA* II. The move from the emphasis on the material stuff of the first half of *PA* II to the polymorphism of movements and activities of the whole seems to be echoed by something Hans Jonas says when examining biological foundations of individuality:

> On the one hand the living being is a composite of matter, and at any time its reality totally coincides with its contemporary stuff—that is, with one definite manifold of individual components. On the other hand, it is not identical with this or any such simultaneous total, and its reality is not bound to the assemblage making it up now as this is forever vanishing downstream in the flow of exchange; in this respect it is different from its stuff and not the sum of it. We have thus the case of a substantial entity enjoying a sort of *freedom* with respect to its own substance, an independence from the same matter of which it nonetheless wholly consists (p. 191).

Chapter 5
Finding Fault with Nature

Abstract As might be expected, the discussion of the parts attempts to get at some knowledge of cause (646a8–13). In this effort, it appears as if the movement of the argument keeps getting pushed towards function. There are distinctions to be made within function: we can distinguish common functions of parts, functions that cut across kinds, but we also recognize functions that are particular, functions that are not cross-kind; this follows the general pattern set out in *PA* I (645b25), which indicates a unity to the book. Take teeth as an example. With regard to teeth, the common function lies in the working (*ergasian*) on the food, while some organisms have specialized functions that differ according to kind (*kata genê*).

Much of the argument of the *PA* seems to revolve around not only the final and necessary, but also around the common and particular as it relates to function. We should keep in mind how the differentiation within cause (final versus necessary) is meant to affect our understanding of common and particular functions and vice versa. We have already seen how an animal's working-on food is described as the common function or nature of the mouth. The mouth is described, in the new beginning of *PA* II, as one of the two most necessary parts (*duo ta angkaiotata moria*) because of its role in feeding (655b30). Numerous photographs of different animal mouth morphologies with feeding strategies are used to illuminate the analysis. To oversimplify to some extent, in the context of *PA* II, the emphasis is on necessity, while the emphasis shifts even more to function in *PA* III. In this way, *PA* III continues and fills out the argument of the latter half of book II. *PA* III focuses on functions as they are common, according to kind and those that are particular.

5.1 The Multiple Roles that Teeth Serve

In some animals, the teeth serve as weapons, sometimes in an active (*poein*) or offensive way and sometimes in a passive (*paschein*), defensive manner (661b3). In general, teeth are used to chew, which is the common function among all animals. This is in line with what is said regarding the mouth as one of the parts

J.A. Tipton, *Philosophical Biology in Aristotle's Parts of Animals*, Studies in History and Philosophy of Science 26, DOI 10.1007/978-3-319-01421-0_5,

most necessary above all (655b30). But when animals use teeth for more than one function it is usually for defense (self preservation) or offense. The exception of course is the human animal, where the additional function of the teeth involves, not their use as a weapon, but in speech. The teeth allow for the stream of sound (the vowels) to be chopped up.

> Mankind has teeth beautifully (*kalôs*) suited by nature to their common use (*pros tên koinên chrêsin*)—those in the front sharp in order that they may cut, the molars flat in order that they may grind. And the canines demarcate these from each other, their nature being intermediate between the two; for the intermediate participates in both extremes and the canine teeth are in a way sharp and in a way flat. And it is likewise with the other animals too, those that do not have all their teeth sharp. But is especially for dialectics (*dialekton*) that mankind has teeth such as these and as many as these. For the front teeth contribute greatly to the generation of articulate sounds (661b6–15).

The case of teeth in humans highlights the relationship between the common and more specialized functions of parts. We move from an investigation of human teeth as they perform the common function of all teeth to an account of their function in particular, peculiar human activities.[1] In the above formulation we have an instance where the beautiful and some aspect of the useful or good come together. But given the stated concern of the *PA* we can ask, What is the cause of the structure and arrangement of human teeth? Did food and diet help shape human teeth (see below and 674a2ff.)? Or is it possible that the cause of the arrangement of human teeth rests on their usefulness for dialectics or speech? The mixture of human teeth also seems to allow for indiscriminate feeding, for omnivory. Even if an activity or function is of a common nature—occurring among many or all animals—we can highlight specific differences in the way the parts of the organism perform the common activity.

This discussion of the role of teeth in speech picks up where *PA* II.16 left off. "Now vocal speech is composed," Aristotle says there, "out of articulate sounds and if the tongue were not such as it is nor the lips moist (*hugrôn*), most of these articulate sounds could not be spoken, since some result from pressing of the tongue, others from pursing of the lips. But what sorts of sounds there are and how many, and what their differences (*diaphora*) are, must be learnt from those who study meter" (660a3–8). The end of *PA* II explains the ingredients of logos with the two parts, the tongue and the lips, which have been discussed. These parts, especially the lips, are described in elemental terms; in other words, the lips are characterized by their fluidity. The human lips, Aristotle says, have a double function as well, to protect the teeth and facilitate speech (*logos*) (659b35). With the addition of the discussion of teeth at the beginning of *PA* III, we see that teeth, in addition to the other parts of the mouth, have a role in the *genesis* of the letters and thus, furnishing the necessary conditions for speech.

[1]One wonders if the distinction between offensive (*poein*) and defensive (*paschein*) that was applied to the teeth functioning as weapons in some animals also applies to their functioning in human speech. Is there an offensive and defensive distinction to be drawn in speech?

The discussion of lips, tongue and teeth does not proceed smoothly; it is interrupted in the transition from *PA* II to *PA* III. Why is the flow of the argument interrupted? There are several answers that we could offer: (1) Aristotle indicates something about the importance of this discussion by making it disjunct; (2) the arrangement of the books was done haphazardly; (3) something in the terms of the argument has changed, for example talk of *logos* being replaced by a concern with dialectics or the introduction of desire for pleasure which accompanies food (661a7 ff.); (4) something about the discussion of the subject matter in terms of the elements, such as the fluidity of the lips, necessitates that the discussion of *logos* as a peculiar function of these parts be determined by the elements. As noted above, the mouth in *PA* II is identified as one of the two most necessary parts while the emphasis shifts in *PA* III, more to function.[2] What is necessary now becomes what is the common function. The movement from book II to III could be described as a deepening of the argument that examines how the necessary/final understanding illuminates and is illuminated by the recognition of common, generic and specific functions.

5.2 The More and the Less, Males and Females

In his discussion of teeth functioning as weapons, Aristotle stops to make a parenthetical point about males and females and the more and the less. In this digression, he makes reference to the fact that what we are doing is *historia* (631b28). Such a statement points back to the methodological questions posed in *PA* I.1 and the beginning of *PA* II. How is *historia* related to our attempt to give a causal account of the arrangement of parts? Or does the fact that we are engaged in *historia* indicate that we have temporarily left the concern with causes behind?

The recognition of some sexual dimorphism with regard to the occurrence of some parts prompts a note on the more and the less. "Nature (*phusis*) allots defensive and offensive organic parts," Aristotle observes, "only to those [creatures] alone which have the power (*dunamis*) to make use of them, or allots them in a greater degree, and in the greatest degree to the animal which can use them to the greatest extent (661b30)." Males are stronger and more spirited (*ischuroteron kai thumikoteron*) than females (661b31) and thus they have those parts Aristotle has identified as functioning as weapons.[3] Strength and spiritedness account for why

[2]The transition from necessity to function is not abrupt insofar as function comes to the fore, for example, in the *PA* II discussion of the elephant's trunk.

[3]In commenting on this section, Ogle (1882/1987, n. 4, p. 187) notes "That the males in any armed species are almost invariably furnished with more formidable weapons than the females is a conspicuous fact, as also is their more pugnacious temperament. The temperament is according to Aristotle's views the antecedent of the weapons. But it is more probable that both weapons are temperament are attributable to one common cause; that what that cause is Darwin has shown on his work on sexual selection."

parts used in offense and defense—in this case the horns—are found in males only or in males to a greater degree. The presence of these parts (e.g. horns, stings, spurs) is not a result of necessity. As we shall see, the issue of horns and their relation to the notion that nature does nothing in vain will return.

Turning to strength and spiritedness is an example of examining the cause behind a given part, which is what is said to distinguish the *PA* from *HA* (646a8–13). This suggests that engaging in *historia*, which is exactly what we are said to be doing in this digression on horns, is still to search for causes; if this is so, then this would necessitate that the perceived distinction between Aristotle's *History of Animals* and the other treatises be revised. In the case of horns, parts that act as weapons are explained on one level by an appeal to strength and spiritedness as opposed to necessity. Aristotle recognizes parts that are sexually dimorphic and attempts to give an account of such dimorphism.

The parts that distinguish males and females most clearly are not horns but genitalia (*HA* 486b29ff, 486a8ff, *PA* 650a2ff). There is no real discussion of the genitalia in *PA* because, as we have noted, reproduction is not addressed, it is left for another discussion.

5.3 Fish Teeth and the Elements

If the presence of horns on males of some types of ungulates can be explained by an appeal to strength and spiritedness, the particular structure and arrangement of fish teeth requires a different explanation. The character of fish teeth is meant to prevent the mixing of the *trophê* and water in the gut (662a9). Fish teeth allow for the segregation of elements. *PA* II highlighted the need to combine elements, but as *PA* III demonstrates, sometimes elements should be kept separate. Perhaps this is one more way to see what has changed in the argument between *PA* II and *PA* III; the elements have become the way in which the habitats of organisms are described.

The watery element in which fishes find themselves along with the need to keep elements (earth, water, etc.) separate in feeding helps to explain the arrangement of fish teeth (Fig. 5.1). Aristotle will explore further the idea that food and lifestyle shape an animal's parts. As we have already seen, feeding may be something which all animals partake in, but it does not involve the same parts or strategies; or perhaps it does only in the most general way. The peculiarities of the aquatic life dictate that fish tackle the "common nature" of feeding in a unique way. The various and particular ways in which the animals overcome certain common problems will lead to a generalization regarding the many and the one.

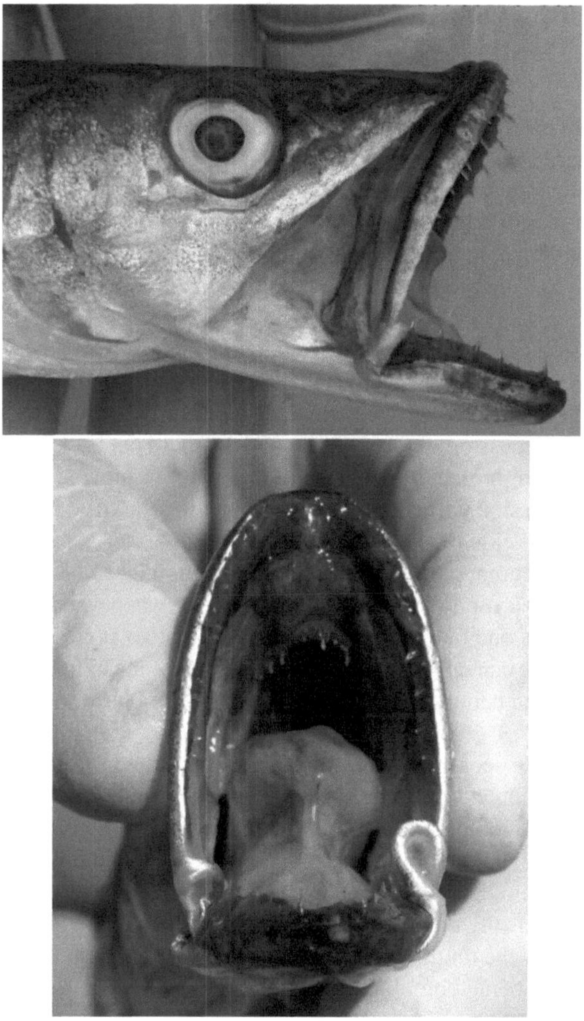

Fig. 5.1 A European hake (*Merluccius merluccius*), with a row of teeth on the roof of its mouth. Aristotle was very impressed with the hake's mouth, likening it to that of a shark (*HA* 600a1). The arrangement of teeth, especially those farther down the throat, allow the fish to take in prey without having to chew and letting water in. This arrangement allows for fish to segregate the elements, keeping the water out. Aristotle notes that the hake hides for a long time since there is a period when it is not caught (*HA* 599b33). Aristotle also remarks that hake, like other demersal ambush predators, hides in the sand or near the bottom waiting for prey (*HA* 628b30)

5.4 Many Into One and One Into Many: The Case of the Mouth

Horns and teeth can be addressed together insofar as they share certain similarities, e.g. they are hard parts which can be used in defense or offense. As the discussion of fish teeth illustrates, teeth are used most commonly in chewing food. The teeth are part of the mouth while also playing a role in nourishing the whole organism; it is a part which points us in the direction of the whole. At this point, the discussion moves from the parts of the mouth to the mouth as a part of some larger whole.

The discussion of the mouth is very similar to the discussion of the parts of the mouth (i.e. lips, tongue, teeth, see 660a1 ff.). We can note that the mouth is a part that can be further broken down into recognizable parts such as lips, tongue and teeth. It appears as if the discussion of the mouth is an attempt to put back together into a one that which had been broken down into discrete parts in previous discussions. This aspect of the discussion will culminate in recognizing the fusion of several parts of the mouth, as in the case of birds.

And like the different parts that make it up, the mouth can be used in the service of defense or attack, or for speech. It appears that whenever something such as teeth has more than one function, there appears to be two alternatives for that more peculiar function; it can either function as a weapon, or it can function in speech, as is the case of human beings. While human beings do not have weapons *per se*, they do have speech.

We have already seen how parts often have a common function (e.g. the lips preserve the teeth) in addition to a specialized or particular function (e.g. lips play a role in the production of speech). With the mouth, we are given an example of a function that occupies some middle ground between the two poles of common and particular. The common function of the mouth lies in chewing food, Aristotle refers to this as the universally common (*pantoôn koinon*) (662a20). Particular or private (*idios*) functions of the mouth include its use as a weapon or as a means of speech. Respiration, while being common (*koinon*), is not a universal (*ou pantos*) function of the mouth.

In some cases, these different functions or activities—reduction of food, respiration, and peculiar functions such as speech—are present in one part, namely, the mouth in this case. A suite of characters or functions is packed into a given part which itself varies and in that variation allows for a multiplicity of functions or activities of the whole. As Aristotle formulates it, "nature has collected these uses together in one, producing a differentiation of this part for the differences of its operation" (*tes ergasias diaphoras*) (662a23). The mouth is a part, it is also a kind of whole that includes parts like lips and tongue. It is also a part that includes many functions, as well as other parts. It has varied functions and varies between kinds and thus allows for a multiplicity of functions or activities of the whole. This could be described as a many becoming a one, which then becomes a many on a

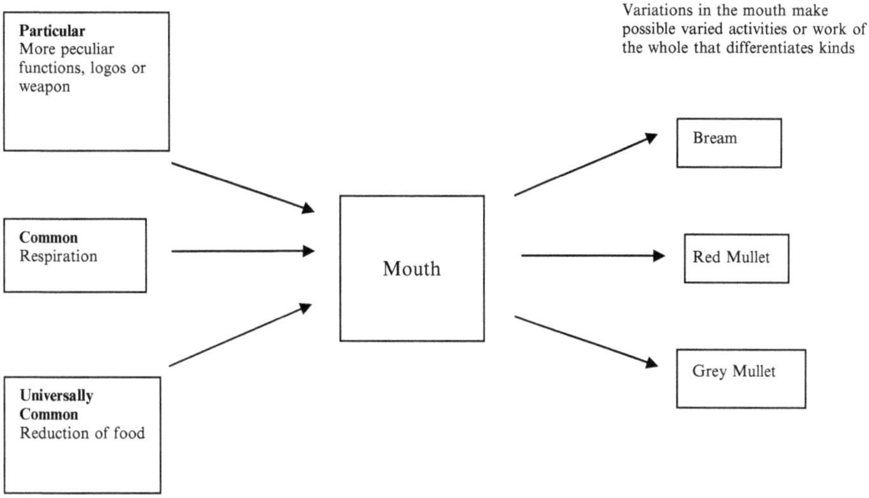

Fig. 5.2 The many functions that are embedded in one part, which in turn allows for a variety of lifestyles

different level (see Fig. 5.2). Such a formulation may make the ontological concerns of the discussion more apparent. This discussion also seems to refer back to certain observations regarding polymorphic actions (645b15; 656a4 ff.).

On the one hand, *diaphora* is now applied to the varied functions of a given part; on the other hand, function or the activity of an organism can differentiate it from other kinds. *Diaphora* can be discovered in the functions of the parts, which can in turn differentiate species or kinds (Figs. 5.3 and 5.4). Just as *diaphora* is used to describe both the different parts and functions of those parts, as well as the workings of the whole organism, *ergon* can also be applied to these different levels, to the parts and to the whole. How the use of *ergon* and the applicability of *diaphora* mirror one another is further illustrated by Aristotle's discussion of the various beaks or bills of birds (Fig. 5.5).

5.5 The Useful: An Analysis of Kinds of Birds with Special Emphasis on Body Plan and Diet; or the Cause of the Heterogeneity of Birds

Mouths do not necessarily have to be comprised from discrete parts like lips and teeth. The bill of a bird is a good example of this insofar as it is a fusion of two parts, the lips and teeth (e.g. 662a34). Beaks are differentiated (*diapherei*) according to use (*chresmos*) and assistance (*boethias*) (662b); every bird has a beak that is

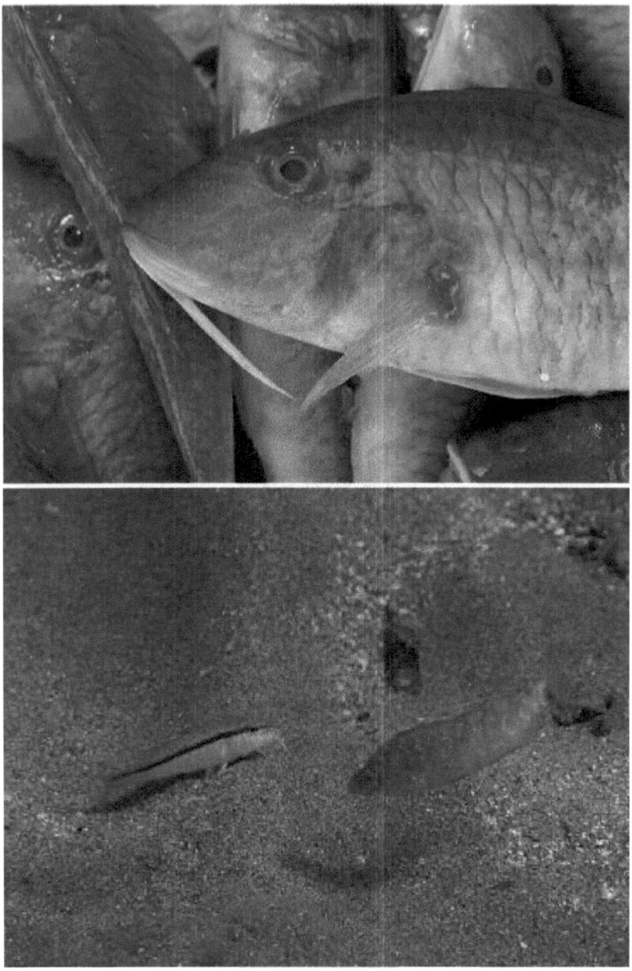

Fig. 5.3 The prominent chin barbels on a red mullet in the fish market (*above*). Note the chin barbels extended as *Mullus surmuletus* forages for food (*below*). *Symphodus cinereus* accompanies the striped red mullet in a search of the sandy bottom. Aristotle was very interested in the feeding behavior of the red mullet, especially its co-feeding with other species (see Tipton 2008)

useful for its particular manner of life (*bios*) (662b7; Fig. 5.6). As Darwin employs the finches, Aristotle gives examples of differences in bill morphology based on some niche exploitation; woodpeckers (Fig. 5.7) have beaks that differ from birds that feed on plants and live by marshes (662b7). Raptors have a curved beak, and sharp talons, useful in hunting prey (Fig. 5.8) and tearing the flesh of prey items that are sometimes quite large (Fig. 5.9). Small birds have beaks useful for their particular feeding habits. The broad (*platton*) beak that allows some birds to dig

Fig. 5.4 Grey mullet (*Mugil cephalus*) have rough patches on their lips (*above*) that they use to scrape algae off of rocks for food (*below*). Aristotle notes that the grey mullet is one of the only fishes that eats no meat (*HA* 591a20)

for roots is analogous to the broad snout that allows pigs to root around (662b12).[4] The morphology between broad bill and snout is similar because of their similar habits with regard to food; the resemblance of the pig snout to the bill of some birds seems to be an example of what Aristotle calls analogy (644a22).

---

[4]On this point, Lnnox (1987) notes that "Different sorts of birds may have beaks of differing length, width, hue, hardness, curvature. It is these sorts of differences, throughout all the differentiae of the general kind, which differentiate one form of bird from another. Along any parameter one may choose, then, each organ will differ only by shades and degrees from one kind to the next" (p. 342).

Fig. 5.5 An egret patrolling the lagoon waters. The long neck and beak, which is like the line and hook of a fishing rod (*PA* 693a20) allow egrets to get their food from relatively deep water

Aristotle discusses the fact that certain parts are present for-the-sake-of certain activities or certain modes of life (662b10): some birds have broad beaks which enable them to dig for roots easily. The variations between beaks or characters in general allow for the habitation of different and varying niches. We usually think about parts as being in some arrangement that is for the sake of the whole, but here

Fig. 5.6 Great white pelican (*Pelicanus onocratulus*), common in and around Lesvos, Greece. In the *HA*, the pelican is said to swallow mussels and "after concocting them in the region before the stomach they vomit them up so that while they are open they may extract the meat and eat it" (*HA* 614b26–31). Pelicans are also a good example of a certain relationship between legs and feet in birds: "In some cases nature provides length for the legs. In some, however, instead of doing these things it fills in gaps in their feet. And it is on account of this that the swimming birds are necessarily web-footed" (694a30–694b3). Aristotle draws a surprising, to my mind, analogy between web feet and fish fins: "it is on account of the better that they have such feet, for the sake of their way of life—in order that, since they live in water where wings are useless, they will have feet that are useful for swimming. For they become oars for sailing just as do the fins of fish; and this is why if the fins of fish or the filling between toes of the waterfowl deteriorate, they are no longer able to swim" (694b5–11). The web feet of the pelican are to be contrasted with the individual toes of a long-legged bird such as a heron

the parts are said to be for-the-sake-of some activity. Birds are wholes but it is their varying parts that allow them to engage in varying activities. Different forms can have different lifestyles, which are a reflection of their varying parts.

In giving examples of differentiated beaks, Aristotle provides us with one answer to the question of why parts are a certain way. The reference to the different modes

Fig. 5.7 A red-bellied woodpecker of North America is similar in many ways to the spotted woodpeckers (*Dendrocopus* spp.) of the Aegean, which are said to have a strong, hard beak serviceable to its mode of life (662b7). Aristotle says that "the woodpecker does not sit on the ground. It pecks at the trees for the grubs and sknipes, to make them come out. For after they have come out it collects them with its tongue; it has a broad and large one. And it walks fast on the trees in any position, even upside down, like the geckos. It has claws better fitted by nature than the jackdaws for security in perching on the trees, for it walks by fixing them in" (*HA* 614b1–8)

of life and the corresponding morphological variations in beaks seem to be part of a causal account. True, the discussion looks like a cataloging of parts and the lifestyles of animals with those parts; however, if this were simply a cataloging, then it would be difficult to differentiate it from the *HA* (646a8), unless it is also giving the cause in cataloging life history strategies. Suggesting that lifestyle maps onto morphology of beak may be part of a causal account. Aristotle notes that the character of an organism's food has the power to determine morphology (674a28). The discussion of the birds seems to support a similar idea. I believe we can confidently assert that the discussion is grounded in the useful (*chrêsimos*); Aristotle looks at the useful in trying to determine cause (cf. 648a16, 659a20). Looking at the useful requires that one examine the organism in its environment. In other words, some part or behavior is useful only when viewed in relation to its surroundings. An examination of life history characteristics is, then, in the service of examining the useful, the good, as a cause.

The observation about bill morphology and body plan of birds that struggle to survive in different ways is powerful and instructive, but perhaps this is a product of reading Aristotle with Darwin in mind. The Darwin finches are often given as an example of the process of speciation and geographic isolation. These birds are all derived from a common ancestor yet they have distinct beaks that allow them

Fig. 5.8 The Osprey or Sea-Eagle is a powerful bird of prey. Aristotle says that "their neck is large and thick, feathers curved, rump broad. They dwell by sea and by coastal headlands" (*HA* 614a4). The osprey (*bottom left*) had just snatched a fish (*arrow*) from a nearby estuary while another one (*bottom right*) struggles with a big catch. Aristotle notes that raptors are impressive predators and have the parts and makeup to succeed in many different habitats. The mention of curved feathers refers, I suspect, to the swept back wings of the osprey in flight (*top right*). All of these photos illustrate Aristotle's suggestion that "the bodies of the taloned birds—excepting the wings—are small on account of the nourishment being used up in their weapons and their defense" (*PA* 694a9–11). There is an allotment of earthen material that is used up in the formation of the claws and beaks (*PA* 694a22–28)

to exploit different resources. For example, one has a parrot-like beak for eating fruit, while others have insect-eating bills, each adapted to a different size class of insect. In bringing in the issue of Darwin, I do not mean to suggest that the two views—the modern evolutionary account and the account Aristotle is offering in the *PA*—are perfectly compatible.[5] But I do think that such a comparison helps the modern reader understand what Aristotle might be doing in the context of trying to determine causes of animals and their parts.

[5]For a recent debate among biologists see Ghiselin (1985) and Mayr (1982). In the context of discussing the omentum, Kullman (1985) makes the following provocative suggestion: "The occurrence of this type in biology shows that Aristotle in his anatomic research arrived at the conclusion that there is anything but a comprehensive finality in animal bodies. It is only through a secondary or tertiary process that an expedient function comes into being. If Aristotle had gone on with these deductions logically, he would have reached the theory of evolution by natural selection" (p. 174).

Fig. 5.9 Here a bald eagle on the Skagit River in Washington state uses its talons and sharp, curved beak to tear the flesh of a king salmon while perched on a partially submerged log. Aristotle remarks that "all the crook-taloned are flesh-eating and cannot swallow grain even if fed by hand" (*HA* 592a29). Such taloned birds do not perch on rocks because it is a hindrance to the curved talons (*HA* 619b8). All of the crook-taloned birds are solitary (*HA* 488a4) (Photo by Zach Tipton)

On the other hand, one could argue for something like a creationist model by pointing to Aristotle's formulation that "nature does nothing in vain." In fact, such a notion appears to be at work in the opening of *PA* III when we are reminded that nature never makes (*poiein*) anything for naught or superfluously (*periergos*) (661b24). In giving concrete examples, however, like the different spaces that differently shaped birds exploit, Aristotle does not fall back on the notion of a demiurgic nature. That leads one to suspect either that such a view is simply assumed or we have gone beyond it. What is unambiguous is Aristotle's reliance on describing the particular beaks as useful (*chresmos*) in the struggle for existence. Determining what is responsible for this usefulness, this match between morphology and lifestyle, directs us to the search for causes in the arrangement of animal parts.

The status of Aristotle's commitment to the notion of a nature that makes nothing in vain is not clear at this point. However, that notion is undermined by Aristotle's accounts of horns.[6]

5.6 The Useless and Nature Doing Nothing in Vain: The Case of Horns

The subject of horn is first introduced in *PA* III in the context of "the more and the less" (662b27ff.) and a discussion of parts used for defence. Polydactylous animals have no horns, they possess other means of defense or assistance. There is a relationship between horns and hoofs; they are made out of the same material (they are made out of the same nature or they have the same nature). Also, there is a finite amount of this material—if nature makes a solid hoof she must remove something from the horns. This explains why cloven hoofed animals have branching or multiple horns; the material is split between the hoofs and horns. Animals with solid hoofs either do not have enough material for horns or, if they do, there is only enough for one.[7]

What we call horns are restricted to the vivipara. Some creatures have so-called horns, but these are not real horns because they do not perform the proper function—

---

[6]Aristotle concludes *PA* III.1 with a brief discussion of the portion between the head and neck that is called the *prosopon* (face). Is Aristotle suggesting that only human beings have this part because it is derivative of the function (*praxis*) it performs, sending forth voice? It is well worth noting that *praxis* is used in this context. Perhaps there is some move from the more general notion of function (*ergon*) to a more idiosyncratic notion like *praxis*, which is only seen in human beings. Man, as animal with logos, is the only creature with a face; humans are the only ones with a part that is designated "face."

Ogle (1882, n. 15, p. 188) notes that "Aristotle does not make the mistake committed by many other writers, of basing man's supremacy on his power of gazing upwards; on the contrary he rightly describes him as looking in front... But that man alone of animals is erect, is repeatedly mentioned by Aristotle as a proof of his superiority. Doubtless the erect position leaving as it does the upper extremities free for skillful manual operations is an important element in man's structure." While I am sympathetic to Ogle's general claim, I would point out that Aristotle does seem to make something of the fact that the parts of human beings are arranged in such a way to view the whole (the cosmos) (656a12).

[7]Humans are polydactylous, their "hoofs" are divided. Is this division the result of a deficiency of material, is it just an accident that humans have divided hands because we don't have enough material for solid or cloven hoofs? Or is it the case that one cannot talk about the genesis of hoofs and hands as being parallel? There is not enough material for either horns or hoofs in the case of humans.

that is, they do not function for-the-sake-of self-defense and attack (662b25). The question becomes, do the animals that have horns that are useless (i.e. deer 663a7) really have horns? The short answer is "no," they have "so-called horns."

Aristotle tells us that the horns of some animals are useless appendages (663a8) and that nature does not give more than one means of defense (663a16–18).[8] We must wonder if horns that do not serve their proper function can be called horns (compare 662b25). Because deer have inadequate horns, they possess speed; in fact, Aristotle tells us that the horns on deer can be harmful (663a11). This situation nurtures doubt on the idea that nature does nothing in vain. Aristotle tells us, deer alone shed their horns to get the advantage of extra lightness: it is a necessity because of their weight (663b12). It is interesting that horns, which are supposed to be for an animal's defense, can work against or contrary to the deer's need for speed. These two characters are working in opposite directions and Aristotle directs the reader's attention to this. If nature had a purpose in mind in doing everything, how could one explain the presence of horns and the need for speed in deer? Aristotle, like Darwin, recognizes the harm that such characters like antlers or horns can have on an individual animal, calling into question the idea that nature has a purpose in mind for everything.

Aristotle presents a doctrine—nature does nothing in vain—and then gives us examples of parts that are not only not useful, but may be at least slightly injurious.[9] It seems unlikely that nature, as some kind of maker or craftsman, would make something so obviously at odds with the standard. Thinking about the functioning of parts allows us to move away from the conception of nature as maker to some other conception of nature.

If nature did nothing in vain, there would be no opportunity to find fault with animals. Why are horns placed where they are? Aristotle claims that Momus in Aesop's fable finds fault with the placement of the horns on bulls. There seems to be several problems with Momus' finding fault. As Aristotle points out, strength is only

---

[8]Regarding the claim about one means of defense for any given element, Ogle (1882, n. 9, p. 190) offers that "This statement, so often made by Aristotle, has much truth in it. For example many insects are protected by the dull tints of their colouring, or their resemblance to vegetable or inanimate objects, rendering them practically invisible to birds. But when they are endowed with some more special means of defence, their colouring is often such as to render them very conspicuous... Facts of similar significance are observable in mammalia. 'Very few male quadrupeds,' says Darwin (*Descent of Man*, ii 257), 'have weapons of two distinct kinds specially adapted for fighting with rival males,' a statement which he proceeds to support by the inverse relation which obtains as a rule between the development of horns and of canine teeth."

[9]While Balme too is skeptical of the formulation, "nature does nothing in vain," he does not emphasize the way in which the argument is undermined by a discussion of things like the uselessness or harm done by horns (e.g. "Aristotle's biology was not essentialist" pp. 299–300 and n. 45.). While I believe he has grounds for being skeptical of some of the formulations, especially the teleological ones, Balme does not attempt to make sense of the movement of the argument of Aristotle's *PA*. I am inclined to think that Balme does not take into account the dialectic between the good and the necessary in the argument, so that the necessary characteristics of the material for example are not incorporated into Balme's analysis (see the discussion on necessary nature below).

one consideration when evaluating their placement; range of movement also has to be considered (663b8). As the fable makes clear, in finding fault in the placement of the horns, Momus was criticizing the craftsmanship of the gods. For Aristotle, finding fault with the parts of animals—as in the case of the horns of deer—is evidence that they are not the product of a divine making, perhaps correcting the notion that nature does nothing in vain. When we discover and examine parts of animals that do not serve a purpose—that are, in a sense, mal-adaptive—we begin to see the way in which luck or chance can play a role in the becoming of animals. The question of the goodness or advantage of certain parts or behaviors would be lost on us as observers if there were not the opportunity to see, in contrast, the useless or harmful. As we shall see in *PA* IV's discussion of the gall bladder and bile, the useless is an important notion in understanding animals and their parts and, perhaps, in our understanding of nature in general.

5.7 Rational Nature Versus Necessary Nature

At 663b22 there is a turn in the presentation of the argument. Aristotle tells us that we have been discussing the horns from the perspective of purpose (*heneka*) or final cause (663b21). We must assume that horns on deer fit into the discussion only by negation, they do not seem to serve a purpose. Aristotle says that now we must speak of the natural necessity (*tês anagkaias phuseôs*) of those things which come from necessity or are put to use by nature for a purpose according to *logos* (663b23).[10]

There is some dichotomy between natural necessity and nature according to reason (see above 640b8–29, 641a25, 642a1). Both formulations, taken together, point to necessity, nature and *logos*. Aristotle discusses the necessary in terms of *genesis*: there is a type of necessity seen in those things that undergo *genesis* and it is *like* (*hosper*) hypothetical necessity (642a7). Hypothetical necessity also involves those things which come to be, those that undergo *genesis*, but is explained by an appeal to the arts and the products or ends that are the result of the arts (639b24). An example of the necessity that is "like" hypothetical necessity would be the impossibility of any animal going without nourishment (642a8): nourishment is necessary if the animal as a whole is to be preserved. This preservation of the animal whole is a kind of *genesis*, a way in which the animal is always becoming, but very different from the generation of an artifact that explains hypothetical necessity.

Three kinds of necessity have been distinguished: necessity in eternal things (639b23), hypothetical necessity (639b25), and what, following Aristotle, we might

[10]This anticipates a very clear formulation of this principle of the relationship between the necessary and the good in *PA* IV: "Some constituents are present for a definite purpose (*heneka tinos*), and then many others are present out of necessity (*ex anagkes*) in consequence of these." One must recognize the necessary in seeing the limits on those parts that have a function, but also determining a cause for those which do not appear to have a function or appear to be harmful.

call "natural necessity" (642a8 and see below for a further discussion). This formulation—natural necessity—is to be distinguished from hypothetical necessity. Such a distinction stresses the notion of nature as opposed to any other process by which things come to be.[11] This combines the notion of necessity and one of the notions of nature. As we have seen, nature can be spoken of in two ways: as matter and as *ousia*, which includes the efficient cause (*kinousa*) and the end (*telos*) (641a25). A form of necessity, natural necessity or genetic necessity, would be associated with the sense of nature that focuses on the matter. In this way, hypothetical necessity seems to be closer to the description of those things which are "put to use by nature for a purpose according to *logos*" (663b23). Evidence for such a view of hypothetical necessity is indicated by the inclusion of the arts in exploring it (639b26). If I am right in making the distinction between hypothetical and natural necessity, the latter might not be explained in terms of ends as definite products or with an appeal to the arts. Natural necessity might be a way in which to explain vestigial organs and parts, those things which are not analyzable in terms of purpose or end, phenomena that Aristotle clearly recognized.[12] The move from hypothetical necessity to that which is "like" hypothetical necessity—natural necessity or genetic necessity—is parallel to the move from the machines or tools (*organon*) of the artisan to the organs of the animal.[13]

The remainder of *PA* III.2 is an exploration or examination of things that exist by this sort of natural or genetic necessity. The first thing discussed in this context is the fact that the larger the animal the greater the quantity of corporeal (*soma*, a word related to body) or "earthy" matter in it (663b25). If the argument turns from a concern with purpose to a concern with genetic necessity, it is not surprising that we are also given a statement regarding methodology: "to study (*theorein*) nature we have to consider the majority of cases, for it is either in what is universal or what happens in the majority of cases that nature's ways are to be found (663b27)."[14] The turn from purpose or teleology to necessity is a turn to the universal. If the concern with purpose or looking at parts in relation to the useful for the animals is a *historia*, the move to the necessary is a shift away from such a perspective. The *theoria* of *phusis* examines the comprehensive and assumes the comprehensive is where we explore things according to nature. But it is not clear that this examination of the

[11]Yet another alternative formula is "genetic necessity" which is indicated by Aristotle's suggestion that "much comes to be owing to necessity" (*polla gar ginetai, hoti anagkê*).

[12]In suggesting that this third kind of necessity, which is "like" hypothetical necessity, might be a way in which to account for parts that do not have a purpose, I am taking a position contrary to Cooper's (1987). While he gives a helpful account of hypothetical, he does not help illuminate how Aristotle uses the notion of necessity to explain parts with no purpose.

[13]Aryeh Kosman (1987) explores the way in which organisms are distinguished from artifacts, suggesting that the elements and parts that compose the former have no being independently from the organisms they constitute.

[14]We can note that in *PA* I, certain things are necessary for the study (*theorein*) of nature (647a27). Now we are theorizing (*theorein*) on nature and necessity in "necessary nature" (*anagkaias phuseôs*) (663b22).

parts of animals is a *theoria* insofar as the inquiry concerns itself with functions of various parts. While the necessary is clear from the perspective of *theoria*, I suspect that it is blind to function. The weaving together of the necessary and the final causes must require a move back and forth between *historia* and *theoria*.

Animals with the surplus of earthy matter can use the resultant residue as a means of defense. Earthy matter courses upwards by necessity (663b35).[15] Nature must do something with this material that is by necessity coming up, so it is utilized to form teeth, tusks, or horns.[16] If female deer do not have horns their teeth should be more fully developed compared to the males; the teeth of both females and males are the same because by nature they are both horned animals; the females have lost their horns because they would not only be useless but dangerous. The horns are less dangerous to the males due to their strength (664a5). It is interesting that the more and the less comes up here again in the context of the horns of deer, but now it is used in relation to the more or the less dangerous; male deer have more horns because it is less dangerous. The question still remains, where does the earthy matter that by necessity courses upwards go in females? Does it just vanish? The stuff that forms horns must be homogenous substance, which is why Aristotle referred to it as earthy matter. But in calling it earthy matter, he is pointing past the homogenous to the elemental. He calls this earthy matter a proper part (*morion* 664a9). In the context of necessary nature, earthy matter is referred to as a proper part. The cause of horns seems to be explained by this upward movement. Only then is it put to use (663b33 ff.) or not, as in the case of deer.

From the perspective of purpose, one could not easily explain the presence of animals that have horns that were of no use, horns that are just "so-called horns." Purpose as a cause does not explain each case. Turning to necessity allows

[15]In the *Origin*, Darwin (2003 (1859)) recognizes a similar phenomenon: "The elder Geoffroy and Goethe propounded, at about the same period, their law of compensation or balancement of growth; or, as Goethe expressed it, 'in order to spend on one side, nature is forced to economise on the other side.' I think this holds true to a certain extent with our domestic productions*: if nourishment flows to one part or organ in excess, it rarely flows, at least in excess to another part*" (p. 147, emphasis added). Darwin is following Goethe's law that "nothing can be added to one part without subtracting from another, and vice versa" (p. 121). Goethe (1988) argues for a "formative force" that drives animal form. Thus he is able to say in the context of this law of economization that: "within these bounds the formative force seems to act in the most wonderful, almost capricious way, but is never able to break out of the circle or leap over it. The formative impulse is given hegemony over a limited but well-supplied kingdom. Governing principles have been laid down for the realm where this impulse will distribute its riches, but to a certain extent it is free to give to each what it will. If it wants to let one have more, it may do so, but not without taking from another. Thus nature can never fall into debt, much less go bankrupt" (p. 121).

[16]Darwin (2004) notes that the teeth and horns stand in an inverse relation: "With ruminants the development of horns generally stands in an inverse relation with that of even moderately developed canine teeth. Thus camels, guanacos, chevrotains, and musk-deer, are hornless, and they have efficient canines; these teeth being always of smaller size in the females than in the males... Male deer and antelopes, on the other hand, possess horns, and they rarely have caning teeth; and these, when present, are always of small size, so that it is doubtful whether they are of any service in their battles" (p. 576).

us to explain the cause of so-called horns on deer. Matter coursing upward is a necessary cause, but not a sufficient condition for the development of useful horns. Material conditions account for the presence of at least slightly injurious parts; the disadvantage of horns is brought about by the character of the material. The advantage or use can be appealed to in searching for cause but necessity or necessary nature seems to lie behind the disadvantageous (cf. 648a16).

5.8 Neck and Esophagus

Aristotle opens *PA* III.3 by saying that the place of the neck, when there is one, is below the head (664a13). This, however, assumes the animal is upright, presupposing a human being. Is a dog's neck below its head? Here, as elsewhere, it seems as if humans are used as a standard.

The neck has several parts (*moria*), all of which seem to be internal: namely, larynx and esophagus. The larynx is present for-the-sake-of breath (*pneumatos*). It is for-the-sake-of the exchange of air (664a18). Air as an element is transformed into *pneumata* when it is incorporated into an organic whole of parts. The esophagus is the passage by which the nourishment (*trophê*) gets included in the body. *Trophê* can include either the wet or the dry or both (664a11). Three of the elements are taken through the mouth but passed on to two different tubes, which again points to the need to separate, and not only combine, the elements.

The larynx seems to include the wind pipe (*arteria*) or the wind pipe seems to be an additional part of the neck. Both larynx and wind pipe serve respiration as well as sound production (*phonê*) (664b1). *Trophê* (including the wet and dry) and air both enter through the mouth; the problem is that there needs to be some segregation. The elements cannot mix at this level so the body keeps them apart. The placement of the windpipe and esophagus within the mouth makes this segregation very difficult. If either the wet or dry elements get into the windpipe, the result is coughing and distress. This must be wondrous to those to hold that animals drink by way of the windpipe (*Timaeus* 7c7). Aristotle has three separate arguments against the people who hold the view that water is taken in by the windpipe: (1) there is no passage leading from the lung into the stomach; (2) there is no doubt were the fluid discharge comes from in cases of vomiting and sea sickness; (3) it is plain that fluid matter which we take in does not collect immediately in the bladder but goes first into the stomach. The windpipe and esophagus are separate and this is validated by empirical evidence.

What had been suggested in the context of fish teeth—that separate paths must be maintained for the separate elements (662a7)—is given more extended treatment. But this is made difficult by the placement of the windpipe (*arteria*); it is situated in such a way as to be susceptible to the intrusion of food. Coughing or choking are bodily indications of the earthy or fluid being where air should be; coughing or choking is an indication, perhaps, of a design flaw. The body combines the elements but also must divide and separate the various elements.

Why would nature put the windpipe where it was in danger of becoming clogged with food? Finding fault with this placement is obviously a Momusian task (663a34). Aristotle tells us that it has to do with the placement of the heart and lungs (665a7). Because the windpipe is susceptible, nature had to contrive a fix, the epiglottis. The mistake that nature makes in the placement of the windpipe has to be corrected by the epiglottis. Nature is not omniscient. The movement and mechanism of the tongue and epiglottis are beautiful, not interfering with the reduction of food. The activity and arrangement of many different parts have to be accomplished in the whole of parts. Explanations for the arrangement of the parts requires appealing to many different causes, including the effect that particular parts have on others:

> For the heart, in which we say the origin of life and of all movement and perception is found, lies in the front and in the middle (for perception and motion are towards what is called the front; in fact it is by this very account that 'front' and 'rear' are defined); and the lung lies where the heart is, i.e. surrounding it, and respiration takes place both on account of this and on account of the origin being present in the heart. Respiration comes about in animals through the windpipe; so since it is necessary that the heart be placed first among things in the front, it is also necessary that the larynx and the windpipe be placed in front of the oesophagus. For while the former extend to the lung and heart, the latter extends into the gut. And generally where nothing greater impedes, what is better and more valuable (*beltion kai timioterôn*) is always, in the case of above and below, present more in things that are above; in the case of front and rear, more in things in front; and in te case of right and left, more in things on the right (*PA* 665a10–26).

Interestingly, the necessary is in line with the best and most honorable. Why should these terms be applied to the placement of the wind pipe based on the prejudice that front is better than back, and above is better than below? These terms are applied to the placement of the organs only after Aristotle has given an account of their cause based on necessity. Their placement is not useful, but more honorable (Fig. 5.10).

5.9 The Viscera, from the Outside to the Inside

It seems as if a good place to make the transition between a discussion of the outside and inside is the mouth. A discussion of the mouth and its parts had introduced the esophagus and the windpipe, thus opening up the inside of animal. Only blooded animals have viscera. Of the viscera the heart and liver are visible as soon as they are formed at all (665a34); these are the first organs visible in a developing embryo (Fig. 5.11).[17]

[17]While praising Aristotle's capacity as a scientific observer, Ogle corrects this conclusion of Aristotle's by remarking that "The heart is not actually the first part to appear in the embryo, but it is the first to enter actively into its functions, contracting in the bird so early as the second day of incubation, and becoming a few hours later rhythmical in its motions" (1888, n.2, p. 193). To me, this is further evidence of the importance that Aristotle put on function and motion in his thinking about biological things.

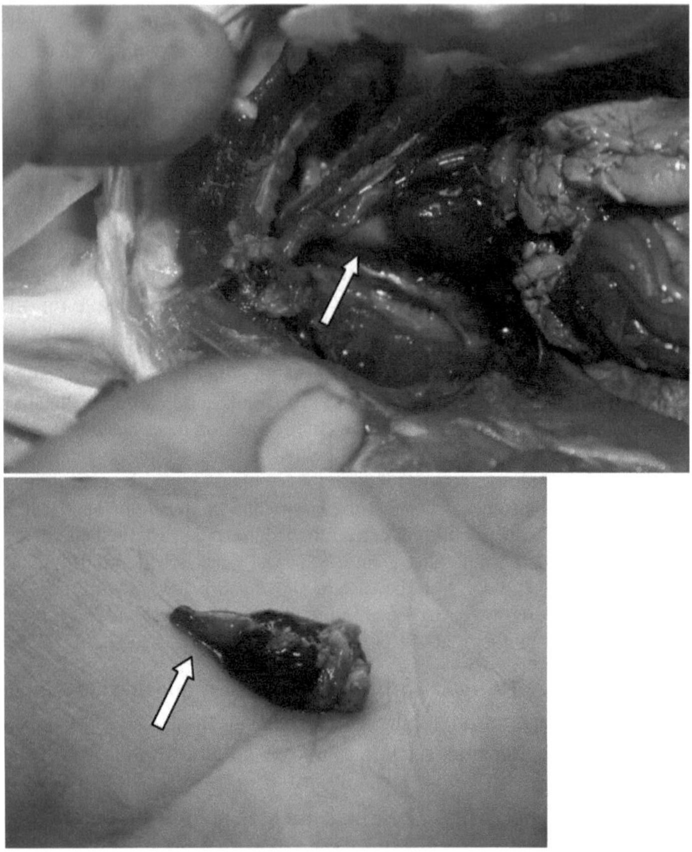

Fig. 5.10 The apex of the heart (*arrow below*) is pointed towards the head, towards the gills which are visible in the photo above. As is clear, the "apex of the heart is sharp and harder than the rest, and lies towards the chest and generally in the front of the body, in order that it should not become cold" (666b). But as he says with regard to fish, the apex is pointed in the direction of movement (666b12)

Aristotle suggests that "just as each animal is equipped with those external parts (*moria*) which are useful for its manner of life (*pros tous bious*) and its motion (*kinesis*), and no two animals require exactly the same ones, so it is with the internal parts: they vary in the various animals" (665b1). As with the beak morphology of birds, internal parts are important in the peculiar way in which individual animals live. Just as in the case of beak morphology, it seems as if the parts are for some motion or activity or manner of life; the formulation of parts for a whole is replaced by the formulation of parts for motion or manner of life. The motion of animals points to something beyond their power of locomotion, hence the concern with passions (*pathê*, e.g. 639a19, 645b5), *action* (praxis, e.g. 645b18) and function

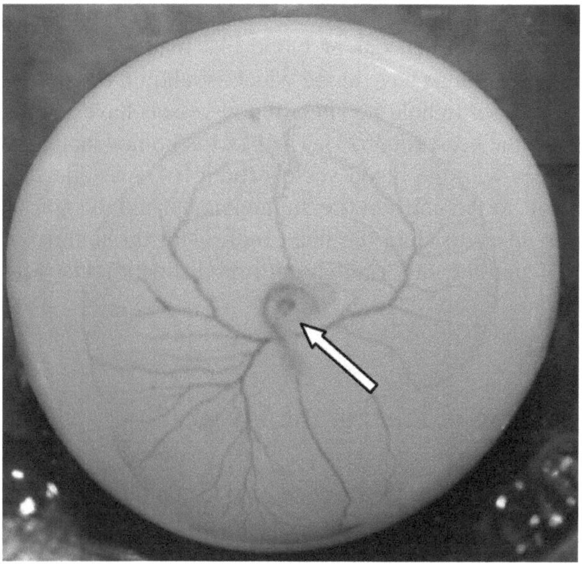

Fig. 5.11 An approximately 72 h chick embryo. The heart (*arrow*) appears like a speck of blood, coming into and out of existence (see GA 734a21 ff.; *HA* 561a4–562a21). "As soon as the blooded animals are constituted," says Aristotle "and while they are extremely small, both heart and liver become visible" (*PA* 665a33–35). It is worth noting William Harvey's description of the heart of chick embryo: "In the middle of this small cloud the throbbing point of blood was so tiny that it disappeared from view on its contraction, to reappear as a red point during its relaxation, and thus between being visible and invisible, or so to speak between existing and not existing, it gave a representation of the heart beat and of the beginning of life" (p. 29). The blood leaves the heart, enveloping the yolk. One can see why Aristotle thought the blood, acting like a stream, flowed out and was responsible for the formation of the viscera. In such observations, one can better understand what Aristotle has in mind when he claims that the "blood is conducted from the heart and into the blood vessels, but not to the heart from elsewhere; for this is an origin and spring of blood, or its first receptacle. These things are more manifest with the help of the dissections and generations" (*PA* 666a6–9)

or work. We believe, and perhaps this is true of the dichotomists, that kind is determined by shape, by *eidos*. *Genos* seems to be in part determined by some aspect of motion or genesis. The core of this argument seems to point back to 641a1–25, to the notion that the shape of the whole is replaced by the activity of the whole. The Democratian emphasis on shape is again shown to be inadequate in the organic world. There seems to be an important connection between whole (that is the whole animal) and manner of life, which seems to point to some phenomenon like *energeia*. This is, in part, why the way an animal looks, or its morphology, is not enough to determine its kind. But does this notion apply also to individuals within a given species?

Variations in the internal parts of animals are crucial for the differing activities. Each of the viscera is formed out of blood-like matter (*hulê*) (665b6).[18] For a blooded animal to be, it must have blood which is a liquid (*hugron*) and there must be, of necessity, a vessel to hold it. These blood vessels have a source (*archê*) and this is by necessity the heart (665b13). The heart is homogenous (*homogenous*) in nature and character with the blood vessels (665b16). Aristotle suggests in *PA* II that the heart points to the uniform (i.e. homogenous) and the non-uniform. In this context he seems to try to align the heart more with the uniform or, at the very least he is not utilizing the uniform/non-uniform distinction that was so prominent in earlier books.

5.10 A Discussion of Each of the Viscera

Aristotle turns his attention to the parts and organs on the inside, the viscera. One might think that it would be sufficient to have one brief discussion of the viscera to cover all animal life—after all, it is just blood and guts. To the casual observer it is clear that the external parts of animals vary between kinds and even among kinds, but not so clear that they internal parts do. Also, as we have already noted, "just as each animal is equipped with those external parts which are useful to it for its manner of life and motion, and no two animals require exactly the same ones, so it is with the internal parts: they are other in the various animals" (665b2). The otherness of the external parts extends to the internal parts and organs, so that the different motions of different animals rely on these differences in the character and organization of the innards.

The heart's connection with motion (e.g. 666a21, 666b18) and the motions of sensation (e.g. 666a12) make it the first of the viscera to be discussed. As Aristotle says, the place in which the heart is situated is the place of primacy and governance (665b19), but the real importance and primacy of the heart must lie in motion. Not only is the heart the source of the motions of animals, but also its own motion suggests to Aristotle that it is like a living creature inside the body that contains it (666b18). Hearts may vary morphologically between the different kinds of animals, but they will all have a motion as long as the animal remains alive. Aristotle gives as an example, fish which "have the sharp part of the heart towards the head, but this is the front; for their movement is in this direction" (666b10).

The variations in the shape and orientation of the heart give rise to variations in the temperament (*êthos*) (667a12). Aristotle suggests, perhaps paradoxically, that a large heart leads to a cowardly temperament, since a heart that is large relative to the size of the animal is like a large room that is difficult to heat, leading to a kind of

[18]Here it is admitted that we can divide up the viscera into distinct parts, which is different from the discussion of the viscera in *PA* II.

coldness that is associated with fear. Certain aspects of temperament were the cause of horns in male deer (661b27 ff.); now we see how certain characteristics of the parts cause certain aspects of temperament.

Many of the viscera can take quite a bit of abuse, in that they can be diseased to a certain extent. However, this is not true of the heart. Aristotle concludes this from some kind of acquaintance with sacrificial victims (667b). The heart cannot withstand serious affections (*pathemeta*). Although sacrificial victims offer a source of dead bodies to examine, one wonders why Aristotle brings up the issue of sacrifice. Sacrifices of all types are usually made to the gods. Aristotle takes the phenomenon of sacrifice in order to learn something about the parts of animals. Aristotle uses an act that is usually performed in the service of the pious in the service of understanding the parts of organisms.

Aristotle can say that the heart is the principle (*archê*) or source of governance; but more concretely it is the source (*archê*) of the blood (666b29) (Fig. 5.11). Because there are two main conveyances for blood—the great blood vessel and the aorta—Aristotle suggests that the heart has to have a kind of doubleness to it (666b30 ff.). The doubleness of the heart is in the form of the multiple pools of blood that supply the two main vessels. As we have already seen, the heart is an unusual part, being between the uniform and the non-uniform (Figs. 4.2 and 4.3). The doubleness or the multiplicity of the heart demonstrates that it is a composite thing, that it is an articulated whole (*diarthrosin*) (667a7). In its motion, the heart is like a living thing encased in a living thing. It is also an articulated whole embedded within an articulated whole. The soul is sometimes the subject of a division at the joints (*Phaedrus* 253d), but here Aristotle recognizes the jointedness of the heart based on the movement of the blood through the two major vessels. The discussion of the heart as a whole, with its movement and articulations, within a whole, cannot but remind us of the discussion of the movements and articulations of the bones that make up the skeletal whole (654a32 ff.).

5.11 Blood Vessels: Limiting the Unlimited

Aristotle's assertion that blood is fluid (*hugron*) (667a19) and as such has no definite boundaries (*GC* 329b30), reminds us of the importance of the elements in the genetic account of the parts. The blood vessels, what Aristotle calls the great vessel and the aorta, provide the container for the boundary-less blood; the blood vessels are for the sake of (*charin*) containing the blood. However, the blood does not remain indefinite indefinitely, because it is the material (*hulê*) out of which the entire body is constructed (668a5). The fluid blood eventually becomes solid (e.g. 668a27). The wet has the potency (*dunamis*) to become the dry, but until that takes place the blood vessels must limit the fluidity.

In his discussion of the heart, Aristotle had given an explanation regarding the duality of the heart and its separate sources or pools of blood (666b28 ff.). This discussion was grounded, so to speak, in the fluidity of the blood and the need for receptacles. In terms of its elemental characteristics, blood is hot (649b25) as well as fluid. The heart is the cause of the blood being hot and fluid (667b26), so even though there are two blood vessels they must come together in this one part.

The discussion of the cause of the arrangement of the blood vessels should simply involve the blood's fluidity and heat. However, this discussion is complicated by several factors, most notably with the introduction of the being-at-work (*energeia*) of the sensory soul (667b20). The soul brings up the question of sensation, making Aristotle formulate the cause of the arrangement of the blood vessels as part of a unity: "Thus we see that because the source of sensation and the source of heat are in one and the same part, the blood must originate from one source too; and because there is this one origin of the blood, the blood vessels also must originate from one source" (667b28). Why is the formulation of the blood's heat and fluidity (667b27) replaced with heat and sensation? In other words, the discussion had concentrated on the elemental characteristics; in adding sensation, and the sensitive soul, the terms of the argument are radically altered. Why is soul an ingredient in the causal account of blood and the blood vessels?

The fact that there are two blood vessels that Aristotle recognizes seems to be related to the bilateral nature of animals: left can be distinguished from right, front from back (667b32). That the blood vessels should be two-parted is supposed to be consistent with this general bilateral character. However, it does seem strange that the doubleness of the internal blood vessels should depend on or be derivative of some external symmetry. The twoness of the blood vessels is unique; these two blood vessels are braided together, uniting the body (668b25). Two parts are twisted and plaited together to form a whole, a one. The blood vessels might provide an image of the plaiting together of the most serious division that must be overcome in the organic whole, that between body and soul.

5.12 The Lung

Any animal that resides on land, any animal or kind going by foot (*pezos*), requires a lung. It is a matter of necessity (*anankaios*) that these kinds of animals have a means by which the heat of the body can be cooled (638b34). The previous discussion of the brain (652a24 ff., 652b5 ff.) makes it clear that the heat of the blood and the heart is excessive and requires some cooling counterbalance. Because of the necessity of the lung, the brain does not appear to be sufficient for cooling in land animals. The brain offers an internal cooling mechanism, while the lung provides a cooling from outside of the body. The lung acts as a window for cooling to be effected by air.

In the context of discussing how the uniform parts are comprised of the elements, Aristotle suggests that the hot, cold, solid and fluid are the principles of the natural elements (*stoichios*) (648b9). Animals that are restricted to the water, with the exception of those like whales and dolphins, have a part that is analogous to the lung; the gills use water as the means by which the fish are cooled. The elements—air, water, earth, fire—and the principles of the elements seem to have returned in the discussion. However, instead of being the material components of the uniform and non-uniform parts, they are part of the process of homeostasis. The lung is for-the-sake-of cooling. In one sense, a part of the body is now acting for-the-sake-of a characteristic of the elemental.

Lungs generally belong to those kinds that reside on land or earth, while gills generally belong to those kinds at home in the water. But there are exceptions, as we have already noted. Dolphins are water animals that "partake of the nature of land-animals to such an extent that the end (*telos*) of life for them lies in their breath (*pneumatos*)" (669a12). The elephant, with its extraordinary trunk, has already been shown to be somewhat intermediate between the solid and liquid environments. Apparently, some organisms are at home between the elements or in a mixed state.

The common function of the lung involves breathing; it is the organ (*organon*) of breathing and the source of its motion lies in the heart (669a14). This common function of the lung is set in contrast to other theories about its function. According to a certain theory (*Timaeus* 70c), the lung is present to provide a cushion for the throbbing of the heart, for the violent motion of the heart. Aristotle explains that the excessive motion of the heart is found only in human beings and is generated by hope and expectation (669a20). Saying that the lung provides a cushion for the heart would only be applicable to humans. It would be a particular function of the lung, rather than a common function. But it is not clear that Aristotle would even recognize the lung's acting as a cushion as a particular function. What is clear is that in dismissing one theory regarding the function of the lungs, Aristotle highlights a distinctively human characteristic, the throbbing of the heart generated by hope and expectation.

There is some relationship between the characteristics of the lung, the heat intrinsic to the animal kind, and the animal's size and posture (669a24 ff.). According to Aristotle, an animal grows to a size that is relative to the amount of heat produced by the heart and blood. Thus, we can infer that larger animals generate a greater degree of heat than small ones. Aristotle goes further and suggests that heat makes the body (*soma*) stand upright (*orthoi*), which explains the posture of human beings (669b 5). Heat seems to be the necessary cause and precondition for man's upright posture (cf. 653a30). We should remember, however, the description of man's upright nature being somehow caused by his ability to gaze upon the whole (656a11), since if heat alone were the cause of our upright posture then all large animals would have an upright posture if there is also a relationship between heat and size. But instead, birds, which are described as not having a lot of heat (669a30), stand upright in a sense, they are bipedal.

5.13 The Duality and Unity of the Body and Organs

After discussing the lung, which had followed a discussion of the heart, and the causes and characteristics of this respiratory organ or its analog, III.7 provides a general introduction for the discussion of the other viscera—spleen, liver, kidneys, bladder and diaphragm. Again, much of the discussion of each of the viscera will revolve around two foci; explanation will be given in terms of the necessary (e.g. 670a23, 670a31, 672a2), but also in terms of the purpose or the "for-the-sake-of" (e.g. 670b25, 672a15).

That some of the viscera appear to be single is a mere appearance. All of the viscera have a double-nature (*diphues*) (669b19). The cause (*aitios*) of the doubleness of the viscera is a result of the double structure of the body; the body can be divided, Aristotle suggests, into upper and lower halves, front and back halves, right and left halves (669b21). The primary doubleness of the body seems to be a result of its having a right and left side, it's being bilaterally symmetrical (670a4). The right and left sides of the body come together to form a unity. However, it is not clear whether the sides are similar (670a4) or are in some kind of opposition (670b20). The body is both a two and a one (670a5–8); this description also applies to the viscera.

5.14 Bladder and Kidneys

It is clear that the discussion of each of the viscera seems to revolve around explaining how they are necessary and what purpose they serve. The exception to this general schema seems to be the kidneys, which are said to be present not of necessity, but rather, to serve a good and beautiful purpose (*eu kai kalos heneken*) (670b23). That purpose, Aristotle tells us, is to help the bladder perform its function (*ergon*) better (670b27). The kidney depends on the presence of the bladder, which is only found in animals with blood in the lung (671a1ff.). A lung that has the characteristic of having blood necessitates that the animal drink a great deal; in fact, the stomach alone cannot concoct the volume of fluid that they require (671a4ff.). Balme understands that Aristotle's pairing or correlating of certain parts is the way in which he discovers causes.[19] Drinking too much for the proper functioning of

[19]Balme notes that "by looking for those characteristics which are regularly associated we may detect their cause" (1987c, p. 86). Cf. Furth's (1987, p. 50) understanding of Balme's thesis. Balme gives five examples in which the cause is intertwined with a discussion of the characteristics (1987c, pp. 86–87). To take just one example, the epiglottis is not present in animals with scales nor those that are feathered because of the dryness of the flesh and skin (Balme 1987c, p. 86; *PA* 664b22); what Balme does not point out is the *cause* given for the epiglottis' being present. The key to understanding the cause of the epiglottis is choking on food or drink, which tells us something about the windpipe (664b30). Coughing or choking are bodily indications of the earthy or fluid being where air should be. The body combines the elements but also must divide and separate the various elements. Aristotle tells us that it has to do with necessity, with the placement of the heart and lungs (665a7). As noted, the epiglottis seems to be part of a finding fault with nature (665a7). The epiglottis is there, Aristotle tells us, because of some deficiency. This also might be evidence of the animal features are explained ultimately in terms of each other (cf. Gotthelf 1987, pp. 169–170). Given the causes stated within the discussion of the characteristic, why should we expect that Aristotle would have completed another treatise articulating causes? Is not that what

the stomach requires the presence of the bladder, which is, in turn, aided by the kidneys. This is a case in which a certain deficiency with regard to the workings of an organ, in this case the stomach, makes it necessary for another organ to be present or be structured a certain way. Does the deficiency of one organ *cause* the presence or morphology of another part or organ? It does when we understand the nature of the relationship between parts of a whole. Suggesting that deficiency is a cause highlights the way in which the whole is the cause of the partness of the parts. We can understand any one part as deficient, but the whole of parts compensates, in a sense, for the particular failings of any one of its members.

The kidneys are not necessary for life *per se*, since they only come about in animals with a certain type of lung, which makes those animals drink a lot of fluid (*tês hugrês*). The Pre-Socratics attempted to explain how certain organs like the stomach and bladder were caused by water flowing through the body (640b12–18); in a sense, this is what Aristotle is suggesting is the cause of the presence, function and shape of the organs associated with the concoction of food and drink. But Aristotle points to water as an element of growth and nutrition, rather than as an element *per se*. In other words, the elements like water are not directly responsible for the origin and arrangement of the parts of animal bodies; the elements have a role in the coming-to-be of parts in that they are ingredient in animal growth and nourishment. This is a way in which one can explain the example of nourishment as illustrating that which is *like* hypothetical necessity as complemented by a certain understanding of material necessity.

If the kidneys are not universally necessary because they are present only in some animals (670b23), the formation of fat in and around the kidneys can be explained by the schema that revolves around the two foci of necessity and purpose. The cause of the presence of fat in the kidneys has to do with the way in which blood is concocted. The end (*telos*) or result of the process of concoction of blood is lard or suet (672a1–5). In animals with kidneys, it is necessary that there be fat. The fluidity of blood is transformed into solid fat by concoction. This process is *like* (*hosper*) what happens when solids (*zeroi*) are subjected to combustion or fluids (*hugroi*) undergo concoction: some part of the heat (*thermotetos morion*) is left behind.

In the case of the presence of the kidneys and the fat surrounding them, Aristotle uses the model of the elements and the interaction among the elements to help show how certain organic processes unfold. However, it seems clear, especially in the case of the explanation of fat in the kidneys, that the appeal to the elements is a heuristic device, instead of an attempted reduction. However, this relationship between the organic and the inorganic is still a question that haunts the argument of the *PA*.

Fat in and around the kidneys can also be understood in terms of purpose. Aristotle explains that fat protects the kidneys and preserves their natural heat (672a14–16). Fat in the kidneys can be explained in terms of the necessary concoction of blood or the preservation and safeguarding of the kidneys.

---

he did in his descriptions? In other words, is it possible that the dividing off of characteristics, of attributes, is also dividing off the causes simultaneously?

5.15 The Diaphragm and Parts Divorced from the Whole

The discussion of the diaphragm should have followed the discussion of the bladder based on the order that was being followed (670b30). As a result, the discussion of the diaphragm is somewhat out of place. This displacement of the discussion of the diaphragm mirrors the content of the discussion; talk of the diaphragm introduces the topic of parts that have been displaced or divorced from the whole.

The displacement of the discussion of the diaphragm makes a certain amount of sense, based on its function. The diaphragm divides the heart and lung from the viscera around the stomach, including the liver and the kidneys (672b9–11), so it is understandable that we would want those things introduced and discussed before turning our attention to the diaphragm. The diaphragm protects the source of the sensory soul (*aisthetikes psuchês*), the heart and/or lungs, from the parts associated with the stomach and the process of concoction. The diaphragm safeguards the heart from the heat generated by concoction of food (672b17–19). This is the reason why some call the diaphragm "*phrenes*" as if it took part in the act of thinking (*phronein*) (672b30–32).

For certain activities of a whole of parts, some of the processes associated with some of the parts have to be isolated so as not to disrupt others. The diaphragm isolates or segregates certain parts for-the-sake-of certain activities of the whole. However, this dividing off of certain parts still takes place within the context of a whole of parts; in other words, the division that is affected by the diaphragm does not compromise the integrity of the whole. As we shall see, parts can be actually removed, thus compromising the whole and its activities.

Aristotle indicates that the diaphragm, while not directly involved in the human ability to think, is a necessary pre-condition for the activity of thought. In addition, the diaphragm plays a role in another uniquely human characteristic—laughter by tickling (673a3–10). Aristotle explains the mechanics of tickling as a result of a motion that is transferred to the diaphragm, warming it and producing laughter. The topic of laughter issues into a dark discussion of the phenomenon seen when men struck on the battlefield appear to laugh (673a11). Laughter that is produced by tickling is different from laughter produced by the telling of a joke or witnessing an Aristophanic comedy. Surely, the "laughter" witnessed on the battlefield when men are struck is an example of the mechanical preconditions of the other types of human laughter.

The people who report such happenings as "laughter" on the battlefield, Aristotle informs us, are at least more believable than those who suggest that a head can speak after it has been cut off. This idea that a head can speak after it has been severed from the body seems to have some support in two Homeric passages (*Illiad* X.457, *Od.* XXII.352). Only in poetry can the head speak without the body. The story of the priest of Zeus *hoplosmios* who apparently identified his killer after losing his head (673a17ff.) must be fiction. Aristotle presents several pieces of evidence as to the impossibility of such phenomena. Once the windpipe (*artêrias*) has been severed, there can be no motion (*kinêseôs*) from the lungs, which is required for speech (673a23–25). A certain motion must be transmitted among the parts of the body in order for speech to occur. Such a motion is impossible once the integrity of the whole has been destroyed. In addition to this kind of "physiological" explanation against the poetic notion that speech is possible from a severed head, Aristotle points to the fact that such things are not seen among the barbarians (*barbarois*) who cut off heads frequently (673a25–27). Aristotle uses the practices of those who do not speak Greek—those who are thought to speak a kind of gibberish, which explains the etymology of the word "barbarian" (*barbaros*)—to show the impossibility of speech from a severed head.

Aristotle discusses sacrifices (667b3) and beheadings in the inquiry into the parts and functions of organic wholes. The parts are determined by the whole, so that a hand severed from a body ceases, in a certain sense, to be a hand. But obviously the removal of certain parts, e.g. the head, has the effect of destroying the unity of the whole in a way that removing a hand or limb does not. The head must belong to what Aristotle calls the "necessary body" (*anagkaiou sômatos*) as opposed to the limbs and such which can be lost (665b21–27). The head looks, in a sense, like an appendage, but it is clearly part of this "necessary body."

Aristotle presents a general overview about the necessity of the becoming of the viscera as well as their purpose: "We have now said what is the purpose for which each of the viscera is present; but also they have been formed of necessity at the inner ends of the blood vessels, because moisture, i.e. moisture of a blood-like nature, must of necessity make its way out there, and, as it sets and solidifies, form the body of the viscera. That, too, is why they are blood-like in nature, and why the body of all of them is similar, though different from that of the other parts" (673a34–b3). It seems as if we have completed our discussion of the viscera, and specifically our inquiry into the necessary nature of the viscera and how they are viewed in their function or purpose; in other words, the viscera have been approached and discussed from the two causes that have dominated most of the book, the "for-the-sake-of" and necessary cause. When a discussion of some of the viscera comes up again within *PA* III.14, we will have to address the question of what this discussion adds.

5.16 Membranes and the Sovereignty of the Heart and Brain

As means of a transition between topics, Aristotle makes reference to the membranes that surround many of the viscera and serve as protection (673b4–12). Within this context of a discussion of the membranes, Aristotle makes a statement about the heart and the brain that appears to contradict, or at least complicate, the picture of the relation between these two parts to each other and to the organism as a whole: "It is well said that, the biggest and strongest membranes are those round the heart (see Fig. 5.12) and the brain, which is natural enough, as it is always the sovereign [part] which has to be protected; therefore the heart and the brain, which have the

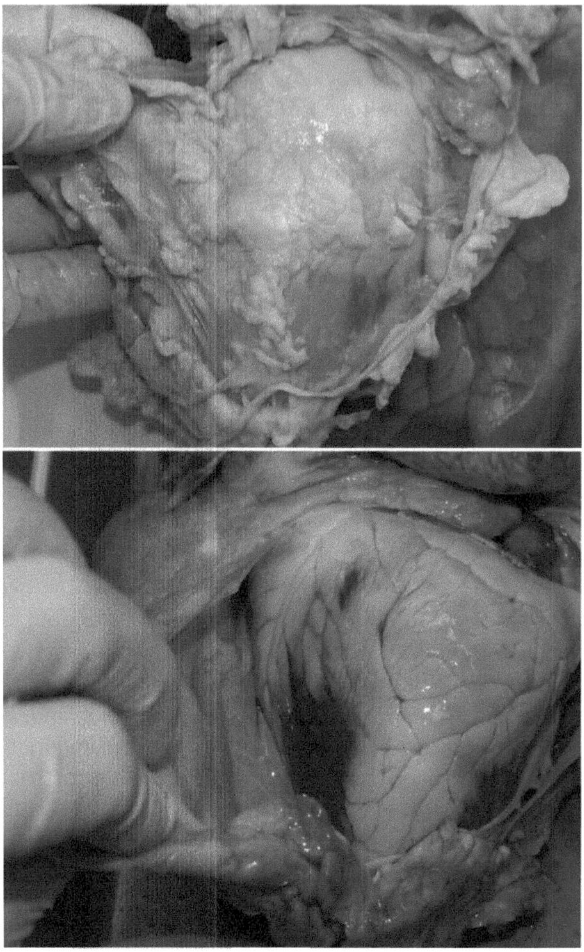

Fig. 5.12 The fatty membrane surrounding the sheep heart. It is rather surprising to see this membrane if you are not expecting it

greatest sovereignty over the living thing require the most protection" (673b8–12). Within the context of introducing the place of membranes, Aristotle admits that the heart and the brain seem to be on a par for the life of the organism. By calling them "sovereign" (*kuria*), Aristotle admits that the parts of the body, even the homogenous parts, are not equally important. Some are more central or critical to the whole of parts.

The admission that both the heart and the brain are sovereign seems to be a correction of the view that the heart is the key part in the organism. Despite suggesting that the brain preserves the animal whole in its nature (652b7), the brain is described mainly as an organ of internal cooling (652a24ff.). In a place the reader does not expect it—a discussion of organ membranes—Aristotle makes the claim regarding the sovereignty of the heart and brain.

5.17 Deficiency or Lack as a Cause

Not all animals have all the viscera. For example, the kidney is not necessary, since it is only present in animals with a certain type of lung. While different kinds of animals may have the same viscera, there may be variations in their form and function: the same viscera can be differentiated (*diaphorein*) in the different kinds of animals (673b15). The earlier explanation of the variation in the parts—"just as each animal is equipped with those external parts which are useful to it for its manner of life and its motion, and no two animals require exactly the same ones, so it is with the internal parts: they vary in the various animals" (665b1–5)—suggests that usefulness is the cause of the variation in the parts of different animals. When it comes to the variation found in the stomach of different kinds of animal, the argument seems to change.

The stomach and gut are necessary for animals in that growth and nourishment are necessary for the maintenance of the organic whole. But the food that is ingested by an animal may be of such a character that it has an effect on the organization of the stomach: "for instance, if the food is thorny and woody and therefore not easy to concoct, in which case the animal has many [stomachs], like the camel" (674a29–31). In the case of the camel, the variation in the morphology of the stomach seems to be caused by the food that the camel eats.

The camel is not the only animal to have multiple stomachs: "every one of the horned animals . . . has several stomachs; and the purpose of them is this: Since the mouth is lacking [deficient] in teeth, the working-on (*ergasias*) the food is deficient; and so one stomach after another receives the food, which is quite untreated when it enters the first stomach, more treated in the next, completely treated in the next, and a smooth pulp in the next" (674b8–13). The cause of the multiple stomachs in some animals is the deficiency of the teeth and mouth; thus, the deficient nature of the function of one part has an effect on another part. It has this effect because both the defective or lacking part as well as those that take up the slack are embedded in a whole of parts.

To illustrate that this is not really isolated to a particular group, Aristotle shows how the same case can be made for birds:

It is owing to the same cause (*aitian*) that the bird kind also differs with respect to the part that is the receptacle of nourishment. For since they too do not fully perform the work (*ergasias*) of the mouth (for they lack teeth)—that is, they have nothing either to cut or to grind nourishment—because of this some birds have, in front of the stomach, what is called the crop in place of the operation of the mouth. Other birds have a broad oesophagus, either a bulky part of it in front of the stomach, in which they store up the unworked nourishment, or some swollen part of the stomach itself; yet others have the stomach itself strong and fleshy in order to be able to store up the nourishment for a long time and to concoct it though it is not ground up. For by means of its potency and heat the nature of the stomach makes up for the deficiency of the mouth (674b18–27).

The deficient character of the mouth in some animals causes the variation seen in other parts like the stomach or esophagus. The argument to this point has seemed to focus on the necessary cause and the final cause, the purpose or the "for-the-sake-of." With the discussion of the deficiency of camel and bird mouths, we are given insight into a new cause, not only in the variation of the parts but in the origin of the parts themselves, such as the paunch (*koilia*) and the net (*kekruphalos*) (674b14).

Not only can a deficiency be the cause of a part's look, but it can also have an effect on the behavior of an organism. In some cases, a defective means of reducing food produces a greater desire for food and a desire that necessarily becomes greater (675a19–24). Aristotle appears to be making room for something like lack or deficiency to have some causal agency. Whatever Momus' intention in criticizing the parts of certain animals, it looks as if understanding such defectiveness is crucial for a proper understanding of life and the causes behind the phenomena of life, more specifically, the causes behind the presence and structure of parts and behaviors of organic beings.

Chapter 6
The Division and Combination of Labor

Abstract Within *PA* IV, two principles, which have been present in various ways in the entire unfolding of the argument, become more fully illuminated: the division of labor principle and the principle of multiple functions. These two principles appear to be connected with the two standards at work throughout the investigation, the good and the necessary.

The division of labor principle assumes a one-to-one relationship between organ or part and function (ala the *Republic's* one man, one job notion). It is founded on some understanding of the best order. The multiple functions principle recognizes that parts often have many functions packed into them—the Delphian knife notion. This principle recognizes necessity in nature. These two principles seem to be incompatible, but there is a necessary movement from the first to the second, entailing a movement from a reliance on the best to a recognition of necessity.

Book IV offers the test to gauge how one understands the whole argument of the *PA*. The reader is confronted with a passage about the placement of mouths in sharks and other pisciverous fish. Many sharks have sub-terminal mouths, which are not at the end of the snout, but oriented downwards. Why is this so? Aristotle first suggests that it appears (*phainetai*) that nature made it this way so as to preserve other animals from them; it takes time for these sharks to roll over to feed on smaller fish and in this time the prey escapes. If indeed Nature makes nothing in vain, this would be a way to account for the placement of this part. However, Aristotle offers two other reasons, which I would suggest do not rely on the making Nature, why the arrangement of the shark's mouth might be positioned in this way: (1) to prevent the shark from overeating and thus saving it from itself; and (2) the shape of the snout made it necessary that the mouth be positioned this way. The movement of this particular passage is (1) from what appears (*phainetai*) to be from the perspective of a broad teleology; (2) to what is best for the animal with this part; (3) to what is necessary given the arrangement and character of other parts of that animal.

J.A. Tipton, *Philosophical Biology in Aristotle's Parts of Animals*, Studies in History and Philosophy of Science 26, DOI 10.1007/978-3-319-01421-0_6,

6.1 The Function of Function-Less Organs

When it comes to organic things, humans want to find causes and purposes. There seems to be a tendency in humans to try to find a purpose in every organ and part that they encounter in animals. This attempt to discover purpose in a part or organ is sometimes misguided, as is the case with the gall bladder. Some would suggest that the nature of the gall bladder is for the sake of sensation (*aesthêsis*); but these people would not be speaking beautifully in making such a connection between the gall bladder and the sensitives, the organs of sensation (676b22). According to Aristotle, the argument presented for the service of the gall bladder in sensation wrongly suggests that it irritates the part of the soul around the liver and that in running free it makes that part of the soul happy (676b24–26). Aristotle argues that this is not correct since not all animals have gall bladders (676b28), but all do have the power of sensation; in fact, sometimes animals of the same group will have a gall bladder, or some other part, that other members of that same group lack. Evidence for this is supplied by the fact that some sheep have enormous gall bladders while others have none (677a2).[1] If the hypothesis about the gall bladder is incorrect, it seems to be a result of a kind of faulty induction: "Whatever an observer has found to be the condition of the individuals he happens to have seen, that he holds is true of every individual throughout the group. The result of this has been a dispute about the whole of the group (*holou tou genous*)" (676b32–36).

If it is incorrect to hypothesize that the gall bladder is somehow for the sake of sensation, it is equally incorrect to assume, as Anaxagoras is said to do, that the gall bladder and its residues are the cause of disease. In this case, according to Aristotle, the Anaxagorean search for cause has erroneously implicated the gall bladder (677a8). That this is not the case is discovered by dissection and examination of the gall bladder and its residues.

We are often wrong when trying to determine the purpose of a part or find the cause of some process in animals. Because he did not know that the gall bladder aids in the digestion of fat, Aristotle suggests that the residue associated with the gall bladder (or the gall bladder/liver complex since they are closely related in the analysis), is just that, a residue without a function in sensation or a causal role in disease. This leads Aristotle to formulate the following principle: "Occasionally nature turns even residues to use and advantage, but that is no reason for trying to discover a purpose in all of them. Some constituents are present for a definite purpose, and then many others are present from necessity" (677a16–19).[2] Just as the "so-called horns" were not of any use to their possessors, making them explainable only by necessity, the presence of bile associated with the gall bladder must be

[1]The fact that Aristotle associates the presence or absence of the gall bladder with particular regions, in this case Naxos and Euboea, might be suggestive of some understanding of geographic variation of a kind. In addition, the fact that sheep are domesticated animals, Aristotle might be pointing out the variation between regions that arises through domestication.

[2]See below the discussion of residues which are emitted by an animal due to some necessity but which serve a good purpose (679a26 ff.).

explained by necessity. A recognition of the necessary keeps us from the misguided attempt to try to find a purpose in every part. On the other hand, this principle that some parts are residues without any function might have prevented Aristotle from further inquiry into the gall bladder and its functions.

Even if the Aristotelian attempt to determine a function for bile fails, it does point us in the direction of a very important part according to Aristotle, the liver. The production of bile around the gall bladder and liver indicates the indispensable or absolutely necessary nature of the liver (677b5). Unlike the heart (677b3), the liver is able to withstand the effects of the bile, which is produced when the blood is not so sweet and healthy (677a26). Bile, while not serving any recognizable function or purpose for the organism does suggest something about the processes associated with the liver in contrast to other viscera. One attempting to understand organic life can learn something even from that which is of no apparent use to the organism itself.

6.2 Necessary Genesis

The case of bile associated with the gall bladder and liver highlight, once again, the parts (or residues) which come to be as a result of some necessary processes and which cannot be adequately explained by an appeal to purpose or final cause. The necessary generation (*genesis ex anagkês*) of the omentum,[3] or the covering of the stomach and intestines, relates to the character of the elements that make it up (677b23).[4] The notion of a genesis out of necessity puts the emphasis on *genesis*, rather than the making of a Nature which makes nothing in vain. When a mixture of the fluid and the solid has the hot added to it, it becomes skin-like and membranous (677b23), like the skin that forms on milk when warmed. In addition, the fatty nutriment that moves through it will add to the fatty character of the omentum (677b25). Given the material character of the omentum, it is not surprising that it is, in part, for the sake of fat deposition.[5] The account of the necessary *genesis* of this part involves an appeal to the elements and the emergent characteristics. It

[3]The fact that only mammals have omentum is something which Aristotle got wrong.

[4]It is perhaps surprising that Aristotle recognized the sacks that surround the heart and lungs and the omentum as things to be puzzled over in light of a concern with the parts of animals. One could imagine a researcher who quickly disposed of these coverings as unimportant, but not Aristotle.

[5]Kullman (1985) includes the omentum in a very interesting discussion that is directly related to much of what I have been arguing. "According to Aristotle," Kullman notes, "the genesis of the thick membrane of the omentum is an inevitable concomitant of the genesis of the intestines . . . afterwards the omentum is given the function to improve digestion. But this function is only secondary. The same applies to the function of preserving the heat, which is done by the fat of the kidney, the protective function of the brows and eyelashes, the protective function of the horns of bovid animals, the protective function of the thick hair on man's head and to the milk of the pregnant female. All these parts did not originally tend towards the end which they now serve" (p. 174).

is as if an account of the material explains what would normally be the efficient cause; some movement or processes that are a result of material interactions are the efficient cause in the case of the omentum.[6] The division between material and efficient causes is lost in this case.

Unlike the bile associated with the liver, which did not have an apparent purpose, the discussion of the omentum can be cast in terms of the useful. Nature takes the product of this necessary genesis of the omentum and presses into the service of concoction: the omentum aids in concoction because it "is fat; fat things are hot, and hot things aid concoction" (677b32). Just as the discussion of the necessary *genesis* of the omentum relied on the elements, the omentum serves its purpose because it is fatty and hot, through its relationship to the elemental hot.[7] Both the account grounded in the necessary and that of the useful appeal to elemental properties.

It is uncertain, however, that outlining the necessary *genesis* of the part provides a causal account of the part. This emerges as a question when Aristotle turns to a discussion of what is called the mesentery (677b36 ff.). He says that what is called the mesentery can be discussed, on the one hand, in terms of necessary genesis (*genesin es anagkês*) (678a4), but on the other hand that the cause (*aitian*) [of that part] in those animals with blood will become apparent (*phaneron*) (678a5). Are *aitian* and *genesis* compatible or equivalent notions in talking about the shape and presence of the parts of animals?

The turn to the mesentery raises another question, first encountered in *PA* I: whether along with hypothetical and simple necessities, there is a third kind that is like hypothetical necessity, highlighted by an organisms' need for food (*trophê*) (642a8). We have already seen this problem of the distinction to be found in necessity (663b22). This issue seems to return in the present context when Aristotle explains that animals must of necessity take in food (*trophê*) from the outside (678a8). Out of this nourishment, the ultimate nutriment (*eschatên trophê*) (e.g. the blood in blooded-animals, 678a10) is generated (*ginesthai*) and distributed to other parts (678a9). The distribution of the blood is similar (analogous?) to the role that roots play in a plants life: "for an animal, the stomach and the intestines correspond to the ground, the place from which the nutriment has to be derived. And the nature (*phusis*) of the mesentery is to contain these vessels, corresponding to roots; they pass through the inside of it. This completes my account of what it is for-the-sake-of (*heneka*) [final cause]" (678a15). We understand the final cause of the mesentery by seeing how the mesentery is to the stomach as roots are to the earth (a similar comparison is made at 650a20). The interest in and emphasis on material, which comes with talk of the *genesis* out of necessity, extends to the elemental level, however

[6]Dudley (2012) might view this as consistent with his claim that "(m)aterial necessity or necessity in accordance with nature (*anagkê kata phusin*), by which a thing moves in accordance with its material nature, is—logically—also absolute necessity" (p. 103).

[7]Lennox (2001) says of the omentum that its role in food preparation "is mechanical in the extreme: being fat, it is hot, and heat aids in digestion. Its function is served simply by dint of its material nature" (p. 291). I want to say, in addition, that the efficient cause of the omentum is simply a matter of its material nature.

metaphorically, when the stomach is compared to the earth or the solid. In the case of plants, the earth is not simply an element, but is the source of nourishment.

The process of converting food into nutriment that can be incorporated into the animal is one way to discuss the generation of animals (*peri tên genesin tôn zôôn*) (678a18). The more fundamental example of the generation of animals is, of course, reproduction. However, a discussion of the parts and organs associated with this more fundamental animal generation is to be put off for another occasion (678a21–26). Animal generation points in two directions: (1) the process of incorporating the raw material for the growth of the animal and (2) reproduction. The *PA* is principally interested in the former but the latter finds its way into the discussion from time to time. How serious is the claim that we can understand *genesis* by examining the parts and processes associated with turning food and nourishment into animal stuff? One could say that the argument revolves around two major points, generation, in the sense of the conversion of food into animal, and preservation. One can understand much about the animated world by examining the ways animals are continuously generated, not in the reproductive sense, but insofar as the individual is preserved and maintained as a whole.

6.3 The Eating of the Bloodless

The analysis of parts like the mesentery highlights the need for an organism to eat, as well as the parts associated with turning food into some "ultimate nutriment" to be incorporated into the body of the organism. While the "ultimate nutriment" is sometimes blood, bloodless animals need some kind of ultimate nutriment, one with parts particular to it. The issue of the need for nourishment also brings up the issue of the pleasure associated with food. Perhaps surprisingly, Aristotle suggests that even the bloodless animals, things like crustaceans and cephalopods, have parts to discriminate pleasant food (678b7 ff.) (Figs. 6.1 and 6.2). Despite the similarity with respect to tongues, some of the parts of animals with blood will be remarkably different or absent in the bloodless animals (e.g. they do not have viscera according to the account). Blooded and bloodless are essentially different (*diaphoran*) (678a27), or, as Aristotle frames it, "the fact that some animals are blooded and some bloodless will be found to be included in the logos that marks off the beingness (*ousian*) of each" (678a34).[8] A privation, bloodless (*anaima*) is a crucial attribute in delimiting kinds and plays an important role in determining the parts that such creatures have.

As a result of their not having blood, these creatures, cephalopods and crustaceans, do not have viscera (678a32); Aristotle must have seen the viscera-like structures in organisms like sepia (Fig. 6.3) and concluded, based on his argument

[8]Lennox (2001) points out that this claim about the bloodless is problematic given what Aristotle says about privative characteristics being part of the defining account of substantial being of a certain class (pp. 294–295). As Lennox points out (p. 295), Aristotle is explicit in recognizing a nutritive fluid in all animals.

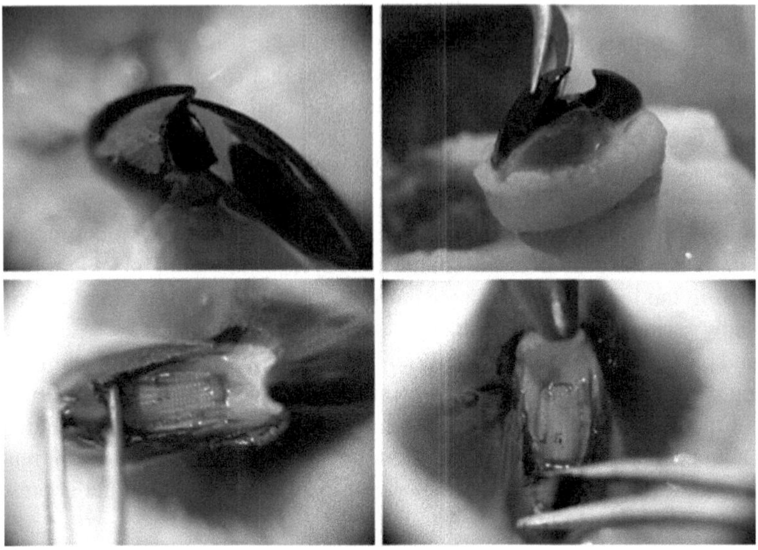

Fig. 6.1 Close up detail of Sepia "teeth" (*top left*), lip-like structure (*top right*) and an organ Aristotle indicates is analogous to a tongue (cf. *HA* 524b4)

Fig. 6.2 One of Lesvos' many kinds of bees. Note the beak like structure with tongue protruding. I am indebted to Hjalmar Dahm of the University of the Aegean's LesvosBee project for showing me his collection

concerning the material make-up of viscera, that the bloodless have organs only analogous to the viscera of the blooded. Bloodless organisms have organs that are not properly viscera because they are not constructed by means of blood. As a result of the argument, the bloodless are said to have a part that is the analog (*to analogon*) of the heart; because of the need for sensitive soul and the cause of life (*to tês zôês aition*) to be situated in a place which rules the parts and the body of the animal (678b3, also 681b15 ff.). This issue of the controlling part, the analog of the heart, prompts a return to the subject of the relationship between the one and many in organisms: "Nature wishes to make this part a one in all, and when she can,

Fig. 6.3 Ink organ of a sepia (*Sepia officianilis*) and "viscera," the *mutis* (*HA* 424b14, *PA* 679a9) (*middle*) and the horny pen or "sword," which is an internal shell of the squid; it is cartilaginous and thin (*PA* 679a20), its consistency midway between bone and "fish-spine" (*HA* 524b25)

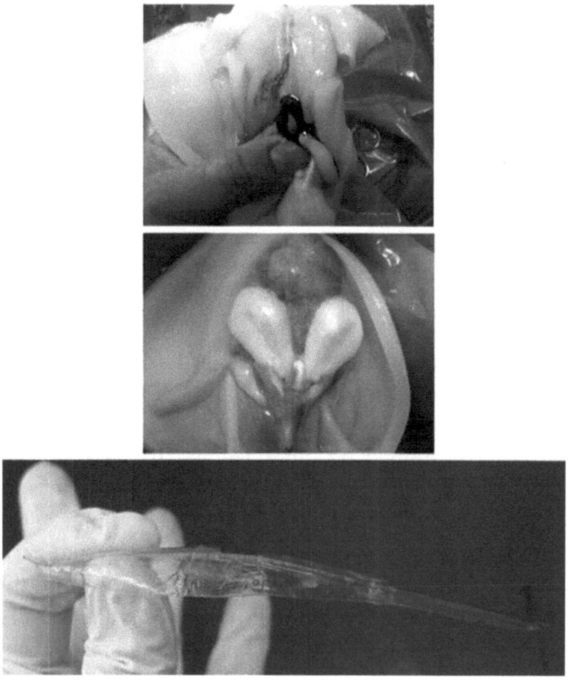

she makes it a one, when she cannot, a many" (682a6). The issue of some kind of potential many-ness of soul applies to things like centipedes which can go on living even when cut up (682a5). This makes some insects like plants (682b30). One can make out of a one a many that continues to be animated.

In addition to a seat for the controlling part, the bloodless also, of necessity (*ex anagnkês*), have the parts associated with food, but these vary (*diapherousi*) according to the places where they take their food (678a5, 681b14). The stomach of the cephalopods (*ta malakostraka*)—including the octopus, sepia and calamari—is similar to that of the birds in that it has a crop continuous with it (678b26) (Fig. 6.3). The cause of the presence and arrangement of the stomach and crop of the cephalopods is the same as the cause of these in birds: "they, like birds, are unable to grind down their food; hence, the crop is placed before the stomach" (678b33–36). The cause is some deficiency; one part, in this case the mouth, does not function well enough, which necessitates the need for another part, the crop. One can note a similarity with regard to the mouths of both birds and cephalopods.

The stinger of certain insects also has a kinship to the arrangement of parts in birds. The stinger of certain insects is a fusion (*suntheton*) of two parts, the tongue and lips (681b11; Fig. 6.2),[9] just as the bird's beak is a fusion of the parts usually

[9]Aristotle's discussion of the parts of insects (e.g. 682a12, 683a30) seems very reminiscent of the discussion of fleas in the *Clouds*. The examination of the parts of animals can be held up as funny;

found in the mouth.[10] The stinger may be a one that is made out of what are usually several parts, but it is similar to the trunk of an elephant in having multiple functions: it is useful for eating (682a20) and for defense (682b35). In this context Aristotle makes a general statement that has a bearing on how we are to understand the relationship between the best, the final cause, and the other causes:

> It is better, where possible, not to have the same instrument for dissimilar uses, but rather the defensive one most sharp, and the one that is to be a tongue spongy and able to draw in nourishment. For where it is possible for two things to be used for two functions without impeding each other, nature is unaccustomed to making things as does the coppersmith who, to economize, makes a spit-and-lampstand; but where this is not possible, nature makes use of the same thing for multiple functions (683a20–26).[11]

In a better world, a world ruled by the best, there would be a division of labor that would result in a certain one to one relationship between organ or part and job (cf. the Delphian-knife, *Politics* I.2). As it stands, what is and would be best is constrained by other factors. It is not a matter of cheapness (*euteleian*), as in the case of the coppersmith who makes a spit-and-lampstand-combination (*obeliskoluchnion*), which is behind the multiple functions packed into some parts. Life is hard and sometimes parts are pressed into performing a number of different services; the struggle for existence does not afford the luxury of combining things for the sake of cheapness, but rather for what is useful and necessary for survival, which can be understood as a kind of economy—the economy of nature.

The principle of the division of labor—one part, one job—must be complemented by the principle, often repeated in the *PA*, that multiple functions are brought together into one part (e.g. 662a24). While the division of labor principle points to the best, the presence of many functions performed by a single part shows that an organic world subject only to the best as cause is not the actual animated world (cf. Lennox 2001, p. 307). The move from the division of labor principle to the multiple functions principle points to the movement from an initial understanding of the perfection and beauty of the animated world to a recognition that the best does not rule simply. The generation of organic things is not like the making of the coppersmith or craftsman; a nature that makes with nothing but the best in view is a fantasy.

---

it is only when one points out the enormous benefit that comes to humans with the knowledge of the examination and study of animals (e.g. raising of livestock, catching fish and other food items, using lions and tigers for sporting events in the Coliseum) does it present itself as serious to the political world. Is this one way in which the study of the organic world is political? In other words, can we take the scene from the *Clouds* to be a warning to couch the investigation in terms of what is useful for the polis?

[10]This, and other instances in *PA* IV, is a case in which the organism at hand is compared to birds. Not only does Aristotle use humans as the standard by which to understand certain things about certain organisms, but when it is appropriate, creatures like birds or elephants (e.g. 683a) can be used as a point of contrast.

[11]On this notion, Ogle notes that: "Here we have a distinct statement of the advantage of division of labour in the animal body; a truth which Milne Edwards thought he was the first to enunciate" (note 15, p. 230. See also Tipton 2001).

6.4 Residues Put to a Good Purpose

Continuing a theme that recurs at many junctures in the *PA*, Aristotle points out the many strategies and parts useful in defense and preservation found in the bloodless creatures, such as the teeth of certain insects (678b20), the changing color of the sepias (679a10), the shells of those with such parts (679b20), and the spines of urchins (679b30) (Fig. 6.4). Even certain habits or habitats are included in the discussion of means of defense, as in the case of the calamari, which is protected by the fact that it lives far out to sea (679a15). In this case, it is not a part that offers protection, but a way of life. All cephalopods have, as a means of self-defense, a peculiar part (*to morion idion*) associated with the discharge of ink (*tholon*) (678b36; Fig. 6.3). The ink is a residue, like the bile of the gall bladder, which is put to a good use, unlike the bile. Fear causes a necessary evacuation of the ink

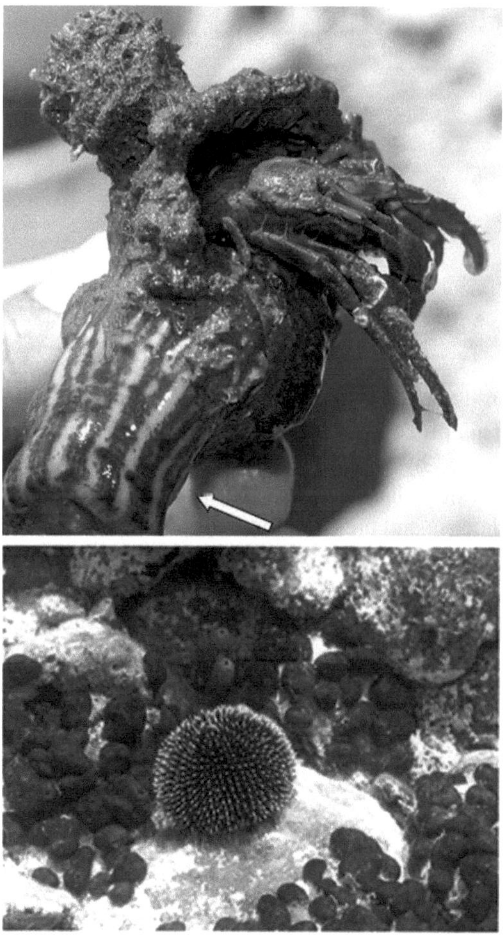

Fig. 6.4 A hermit crab (*Paguridae*) peeking out from under its shell, which originally belonged to a snail. *Pagurus arrosor* is almost always found in symbiosis with a sponge or, in this case, an anemone (*arrow*). Aristotle argues that hermit crabs dualize in looking like crabs but living in shells like the testacea, the hard-shelled animals (*HA* 548a14–21). It is remarkable that this creature, who should be in the same group as the crabs and crayfish, becomes something other by inhabiting a solid shell, a shell like an oyster or murex. Below, an urchin is pictured among a number of sea-squirts

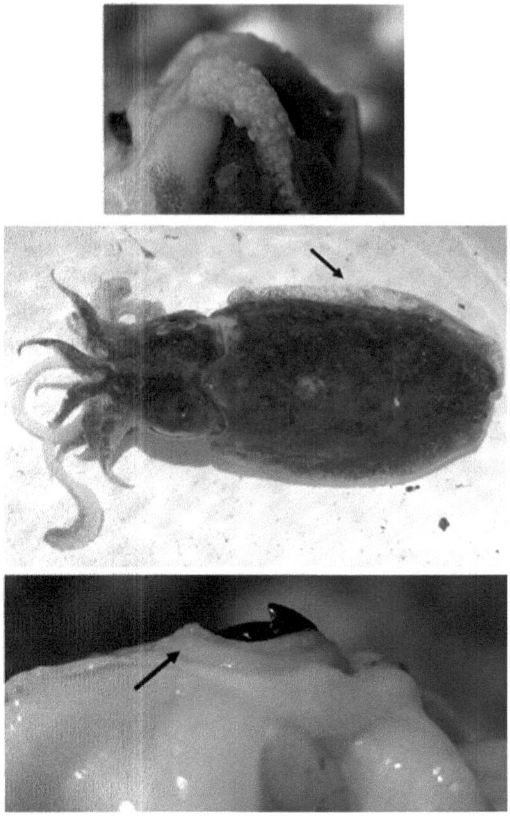

Fig. 6.5 Looking under a sepia, a cephalopod, at the beak-like structure (*bottom*). The fin around the body (*middle*) provides stabilization, much like the tale feathers of a bird as Aristotle suggests. The *arrow* (*bottom*) points out the lip-like structure that protects their "teeth"

in its bag just like some animals discharge their urine in times of fright (679a27). This process, which results in the rapid loss of the residual contents of a bag, is caused by necessity but is turned toward the useful by nature. This is a case where a residue has been incorporated into the functioning of the organism, into its self-defense and preservation. One is tempted to suggest that it is a fruitful accident that the ink of cephalopods is of such use, since this case illustrates the grounding of the useful in the necessary and further demonstrates the constraints and unintended benefits of the character of certain material necessities. The fact that fear, a psychic phenomenon, can cause such physical manifestations demonstrates the intimate connection between soul and body (see the discussion of fear and the changing colors of the chameleon below at 692a20ff.) (Fig. 6.5).

6.5 The Continuum Between the Inanimate and the Animate

The role of fear in the discharge of the ink of the cephalopods reminds us that the organic things are distinguished from the inorganic by possessing a soul. However, in examining the bloodless animals—creatures such as urchins, sponges and sea

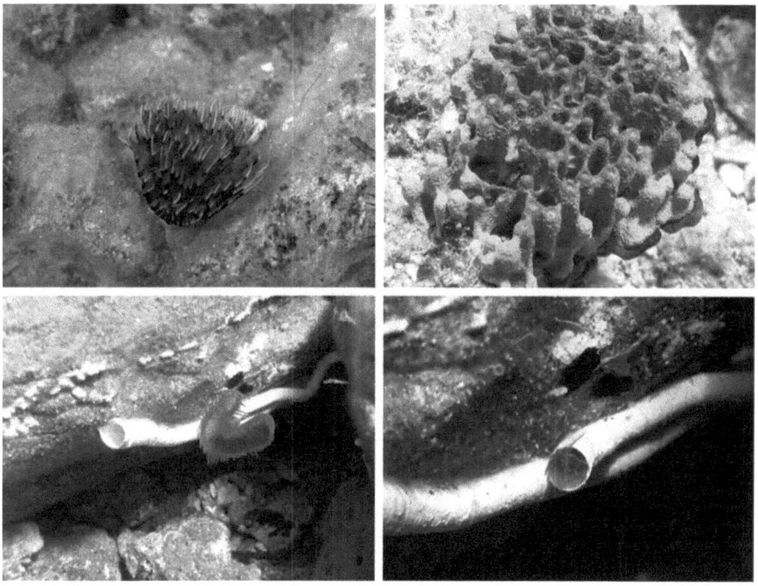

Fig. 6.6 Some plant-like animals including a colony of sponges (*top right*) an anemone (*top left*) and a segmented tube worm, an annelid, protruding from its tube dwelling (*lower left*) and retracted (*lower right*). The anemones "fall outside the divided kinds, and tend in their nature towards both plant and animal. For by being detached and falling upon their nourishment, and by being perceptive of what they fall upon, some of them are animal-like; and further, they use the roughness of their body for self-preservation. But by being incomplete and becoming quickly attached to the rocks, and by having no apparent residue though they have a mouth, they are akin to the kind consisting of plants" (*PA* 681a4–8)

cucumbers—we are struck by their sometimes seeming closer to the inanimate than the animate. At the very least, such creatures seem closer to plants than to animals. It is one thing to see how some part is like the roots of a plant, but it is quite another to see how an animal is almost indistinguishable from a plant. This prompts the formulation of a general principle: "Nature passes in a continuous gradation from the soul-less thing to animals, and on the way there are living things which are not actually animals, with the result that one class is so close to the next that the difference (*diapherein*) appears very small" (681a12, cf. *HA* 8.1). In this, Aristotle is returning to an earlier discussion. *PA* II attempts to give an account of the generation of the animate from the inanimate, by trying to show how the elements can be put together to form the animate whole, thus attempting to show the continuity between the inanimate and the animate; as we saw, the possibility of such an attempted reconciliation was very illuminating, but undermined in the end.

There are many plant-like animals that Aristotle observed, including sponges (Figs. 6.6 and 6.7), ascidians, urchins and some segmented worms (Fig. 6.8). Spurred by a recognition of such plant-like animals, we get a statement of the continuity between organic and inorganic. It is one thing to suggest a kind of continuity of organic things from plant to animal, but extending this to include the

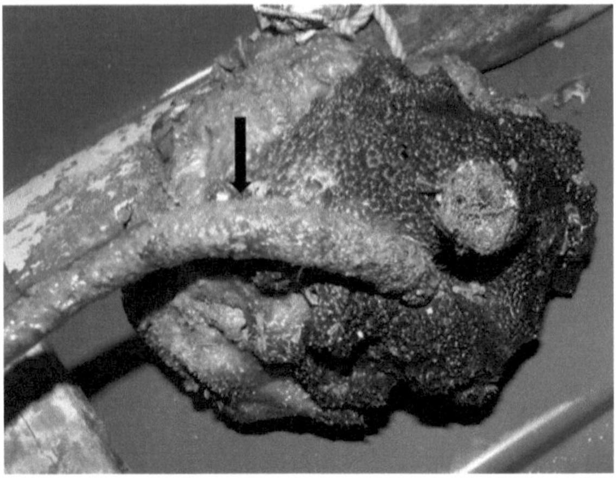

Fig. 6.7 Plant like animals are confusing for a number of reasons. For example, Aristotle indicates that sponges provide homes for many other creatures (*HA* 548b16), which is not something you would associate with an animal. This specimen, collected in the Gulf of Gera, has a tube worm, an annelid, embedded in it (*arrow*), some anemones growing on its surface and some small walking crabs living it its crevices

Fig. 6.8 Some kind of annelid worm, possibly a lugworm; these burrow into the mud in the marshes around the Bay of Kalloni and are used as bait by fishermen

inanimate seems as if it is not supported in the given context. However, examining the elemental processes within the animate whole allow us to make sense of the fact that parts such as the omentum, and other parts, often function by dint of their material, mechanical character.

The creatures that are most plant-like include things like urchins, ascidians, sea anemones and other hard-shelled organisms. Aristotle informs us that there are many *genê* and *eidê* of hard-shelled (*ostrakodermon*) organisms (679b17). In a pattern Aristotle has developed over the course of the investigations, these animals are shown to have parts and ways of defending themselves, as well as parts that are obviously for the purpose of incorporating food into the body ((Fig. 1.3) Figs. 6.9 and 6.10).

Many of these plant-like organisms have shells (that is why Aristotle refers to them as, literally, the "hard-skinned"), which serve as protection, like in the

Fig. 6.9 A purple murex and a murex shell. The two horns are just visible above its muscular foot. In antiquity, purple murex were prized because of the dye that could be made from them, known as Tyrian purple or royal purple. Aristotle describes the part that produces the dye: "The bloom is situated between the mecon and the neck: these are firmly attached to each other. In color it looks like a white membrane, and this membrane is the part people take out: when squeezed it stains and dyes your hand" (*HA* 547a16–19). The name of the part—bloom (*anthos*)—is most commonly said of the bloom of a flower. The so-called tongue of the purple murex can bore through shells of other shellfish (661a21)

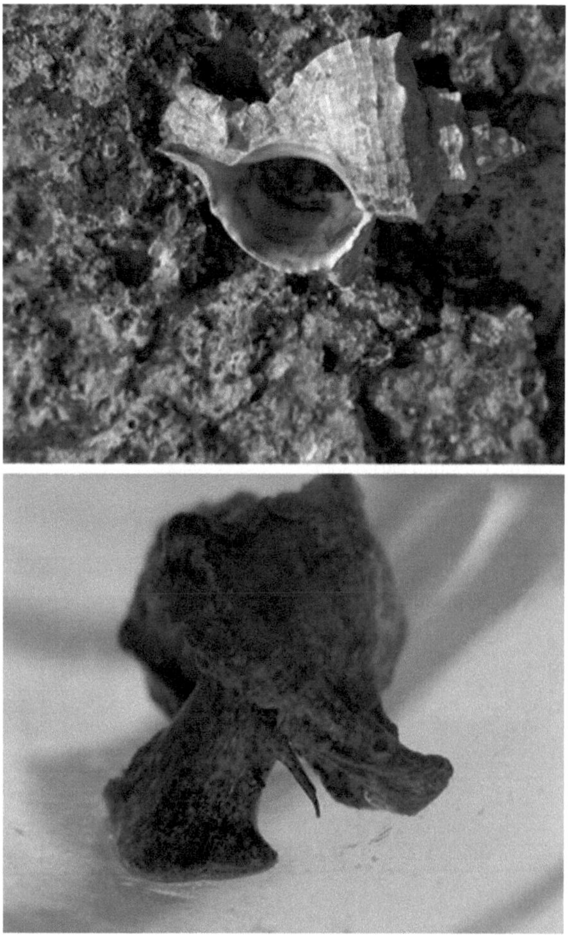

whelks and purpuras (679b19). There exist animals with two coverings, the bivalves (*dithura*), and some with one covering, the univalves (*monothura*). These latter organisms do not have a continuous covering and must adjust accordingly. For example, the univalves (*monothura*) cling to some object, keeping their shells turned toward danger; in this way, the univalves "become (*ginetai*) in a way bivalves in virtue of the borrowed protection afforded by the object to which they cling" (679b25) (Figs. 6.11 and 6.12). Through this strategy, an organism with only one shell that cannot be closed up is provided with a kind of bivalve arrangement. Of the hard-shelled, the one with the most formidable defense is the urchin with its hard shell and spines, which is peculiar (*idion*) (679b29; Fig. 6.13).[12]

[12]While I write this, I have several urchin spines embedded in my foot, serving as a reminder to the truth of Aristotle's observation.

Fig. 6.10 A fan muscle (*Pinna* sp.) almost 1 m in height and in approximately 10 m of water near the modern town of Mytilini. Notice the shells closed more tightly (*bottom*), which occurred after I approached too closely (see *HA* 528a14). Pinna are famous for having a little crab "which acts as its watchman" (Thompson 1947, p. 201). "Pinnae," Aristotle notes, "grow up erect out their byssos in sandy and muddy places. They have inside them a 'pinna-guard'; some of them have a small carid, some a small crab; and if they are deprived of it they quickly perish" (*HA* 547b15). The byssos is a hair-like tuft that anchors the pinna in the mud

With regard to the parts and processes associated with feeding, the sea urchin, for example, has five teeth (680a5, Fig. 6.14).[13] It is, in part, the eating and processing of *trophê* by certain hard-shelled organisms which makes Aristotle puzzled at the boundary between plant and animal. The ascidians (*ta têthua*) have two orifices which are responsible for the uptake of the fluid matter (*hugrotêta*) and discharge of the surplus moisture (681a29, Fig. 6.12). However, this discharge does not appear

[13]This is the so-called "lantern of Aristotle" (*HA* 531a5). Lennox (1983) examines closely the anatomy of the sea urchin to resolve a philological question about a passage in the *History of Animals* (531a3) that draws some analogy between urchins and lanterns. (I would add parenthetically that Lennox's interpretation of the passage would, I believe, be strengthened by noting the discussion of the fact that a shell, in crustaceans and ostracoderms generally, not only can protect the innards, but a faintly burning heat (*PA* 654a8).) So, close inspection of the specific organisms in question can prove very useful in understanding Aristotle's language and, as I would suggest, his thought.

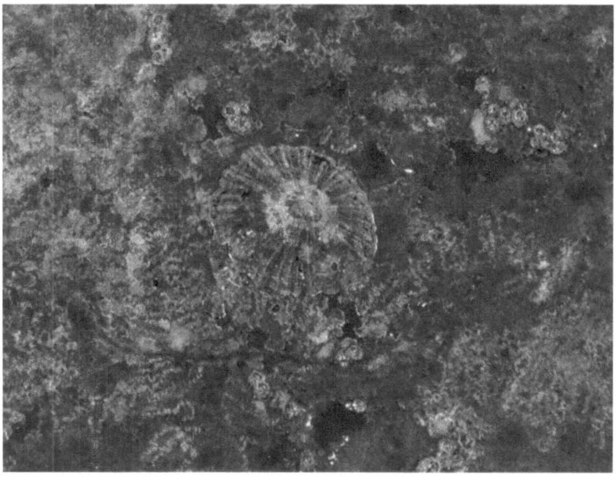

Fig. 6.11 A limpet attached securely to a large rock in the splash zone

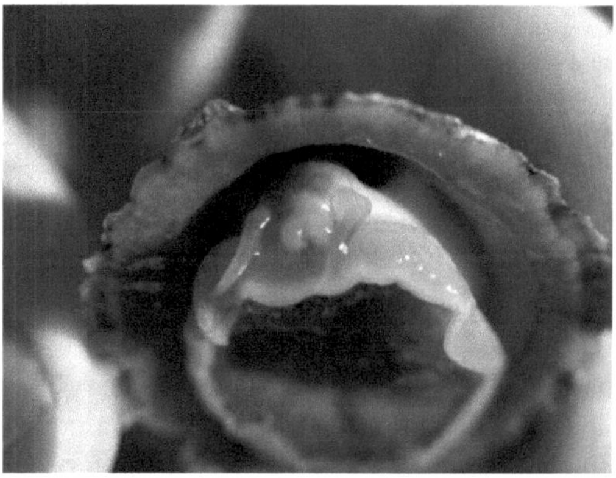

Fig. 6.12 The one shell of a limpet protects a remarkable creature. The mouth, tentacles and mantle are revealed under the limpet's single shell: "they all have a head and horns and mouth and the tongue-like organ, but in the smaller ones these are difficult to see . . . " (*HA* 529a26). Aristotle notes that "even the limpet releases its hold in order to search for food" (*HA* 528b1)

to contain the residue (*perittôma*) produced by other hard-shelled organisms and all other animals (681a30). This lack of residue, which Aristotle sees as a characteristic of plants, leaves in doubt the proper classification of the ascidian.

6.6 The Case of Animal Motion Again

If the hard-shelled organisms bring up the question about the puzzling boundary between the classification of plants and animals, it is perhaps not surprising that examining these kinds of animals also highlights the question of motion. Because many of these organisms are sessile, nature has generally provided them, as we have already seen, with a shell for protection (683b10); however, like the urchin, what might appear to be largely stationary can move with surprising rapidity, using its defensive spines as a means of locomotion (Fig. 6.13). In general, such organisms

Fig. 6.13 An urchin, as Aristotle notes, is covered with spines (*PA* 683b14) which serve, not surprising, as a means of protection but also, perhaps surprisingly, as a means of locomotion. Each spine appears to move independently, allowing the animal to move relatively quickly. The urchin pictured moved across a 2 m long table in less than a minute. Anyone who has snorkeled in the Aegean will have been struck by the daily comings and goings of the urchin populations. This is something Aristotle obviously observed: "the hotness of their nature readies them for greater movement, so as to graze and not remain sedentary. A sign of this is that such sea urchins always have something on their spines, as though frequently in motion; they use their spines for feet" (*PA* 681a6–9) (Fig. 1.2). A part that serves primarily for protection has a more particular function in animal motion

have few parts because of their stationary lifestyle—their bodies are not divided up (*polumeres*) (683b5). Lack of motion in these is a way in which to explain the character of their shell. This observation rests on the recognition of some causal relationship between parts and motions, between the dividing up of the body into parts and an animal's motions[14]: creatures that move about are multi-parted as a matter of necessity, because of their more numerous actions (*praxeis*). The greater the number of motions (*ta kinêtika*) of the animal, the more parts (*organôn*) required (683b6). The relationship between parts and motions is described in terms of cause (*aition*) (683b5)—one causes the other. It is a matter of necessity that more motions require more parts. More parts, with the associated complexity of body plan, allows for more motions. Motion causes parts or can explain the presence and persistence of parts. Simplicity of life—lack of motion—is consistent with simplicity of structure.

Unlike the hard-shelled animals, the soft-shelled ones, the crustaceans, are all capable of motion (683b25) and so have locomotive organs consistent with the different ways of life seen within the group. One could say that the group is united by

[14]In *PA* I, there is a point at which motion is responsible for the division of what looks like a one into many. But that case deals only with the classification, here motion seems to be actually dividing up the organism.

Fig. 6.14 The five teeth of the sea-urchin closed (*top*) and then opening (*bottom*)

the presence of some form of a soft shell, but divided by means of the way in which each moves through the world. For example, the crayfish have a tail "for they swim by propelling themselves with their tails, as if by oars" (684a3, Fig. 6.15). One could also divide the larger genus on the basis of claws. Of the soft-shelled organisms, the shrimp do not have claws, in contrast to the crabs, crayfish and lobsters. Aristotle claims that just as the majority of humans use their right hands, the right claw is the dominant one for crayfish and crabs (684a27). And while lobsters have one dominant claw (Fig. 6.16), it is not on account of its use in grabbing and feeding: lobsters "have claws because they are in the kind that has claws; while they have this part randomly distributed because they are deformed, and do not use it to do what claws are naturally for, but for the sake of locomotion" (684a32–36). Lobsters are members of the soft-shelled group having claws, which seems to be implicitly marked off from those soft-shelled organisms lacking claws, the shrimp. The fact, according to Aristotle, that there is no regularity to which of the lobster's claws is

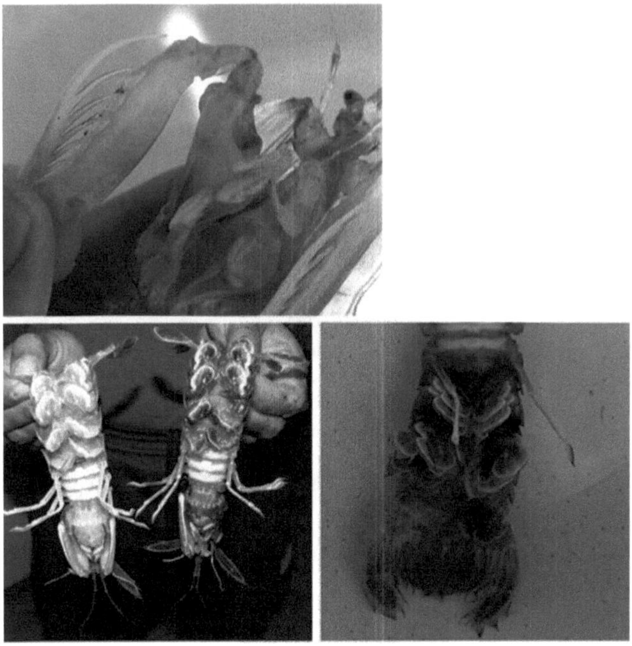

Fig. 6.15 Despite calling it a "mantis shrimp" in English, *Squilla* spp. are to be numbered among the crayfish-like organisms for Aristotle in that they have claws. Of this group in general, Aristotle says that they "have their hard shelly part outside, taking the place of skin, and their fleshy part inside. The under part of them is somewhat laminated; this is where the female deposits her eggs" (*HA* 525b13–14). Mantis shrimp can have claws, says Aristotle, because the material was not used up in the formation of feet (684a15–17). The tail has paddle-like structures that help explain why Aristotle would suggest the tails of these creatures are like oars (684a3)

larger and dominant, suggests to him that it is defective in terms of its natural use and instead pressed into the service of walking.[15] In another context, one could imagine that these claws might be slightly injurious. However, given the morphological affinity of and the spatial continuity between feet and claws, Aristotle might rather conclude that they are inadequate feet of a sort.

6.7 An Organism Bent Over, the Case of the Cephalopods

The motion of cephalopods (*malakioi*) illuminates the somewhat strange relationship between the head and feet, which is reflected in our word, but also in Aristotle's description of their main external parts, consisting in the trunk and the head with

[15]See Lennox (2001, p. 310) for concerns about the argument regarding the use of the lobster claws for locomotion.

Fig. 6.16 While the right claw is always the stronger, dominate one in crabs, the difference in claws in the lobster is "random" in being either the right or the left one. But this account in the *PA* is a little hard to square with the one in the *HA*. There Aristotle saws that the structure of the claws is irregular: "in the *right* one the broad tip is elongated and thin, whereas the *left* one is thick and rounded. Each is divided at the tip like a jaw and has teeth above and below; in the *right* one these teeth are small and saw-like; in the *left* one those at the tip are saw-like, whereas those inside are molar-shaped" (*HA* 526a15). This is a perfect description of those pictured. I am not sure what to make of the differing accounts. Perhaps Aristotle became acquainted with the fact that the dominant claw can vary after committing to the *HA* account

feet around it (684b8) (Fig. 6.17). Cephalopods look as if they have been bent in half when compared to humans and quadrupeds. This comparison is a case where Aristotle takes humans and quadrupeds as same in order to demonstrate the otherness of the arrangement of cephalopod parts. The fact that humans are upright while quadrupeds are not is disregarded; both have head and feet at opposite extremes in contrast with cephalopods. The arrangement of the parts of the cephalapod entails that its residue leaves close to the mouth: the entrance for nourishment and the exit for residue are very close.

The arrangement of feet around the head has some role in the locomotion of these organisms. The legs of sepias are used in swimming and the body fin of these organisms is like the tail feathers in birds or the tail fin in fishes (685b17) (Figs. 6.5 and 6.19). The resemblance between tail feathers, body fin of sepias and caudal fin is not a resemblance of form, but of function; this analogy is obviously

Fig. 6.17 An octopus in approximately 4 m of water. The octopus can change colors, making it look like the substrate, but turning darker when frightened, as the bottom photo captures (*HA* 622a9)

based on function and not the look of the thing. In cephalopods, feet and fin combine for locomotion. With the swimming and walking of the octopus, this organism has two modes of locomotion, unlike those cephalopods which use their legs for swimming only. In addition, Aristotle talks of the function of these feet in anchoring and defense (685a30). The octopus in Fig. 6.17 had its legs anchored in the sand when I approached it; at that point, it lifted itself from the sand and onto the sea floor (Figs. 6.17 and 6.18). The suckers that all octopus have on their feet usually occur in two rows (Fig. 6.20); however, there is a certain kind of octopus with only one row "not because it is best (*beltiston*) but because of necessity on account of the peculiar logos of their substantial being (*dia ton idion logon tês ousias*)" (685b15). The arrangement of the parts of animals is not always in line with what is best.

What is particularly interesting about the discussion of the feet of octopuses is what is not included. The male octopus uses a modified foot as an instrument for sexual reproduction (*HA* 524a8–10). The reproductive function of the specialized foot (the hectocotylus arm) is not discussed in the *PA* although Aristotle was

Fig. 6.18 Aristotle notes that the octopus "swims obliquely in the direction of the so-called head, stretching out its feet; and by swimming in this way it can see forwards (since its eyes are on *top*), while its mouth is at the rear" (*HA* 524a12–13)

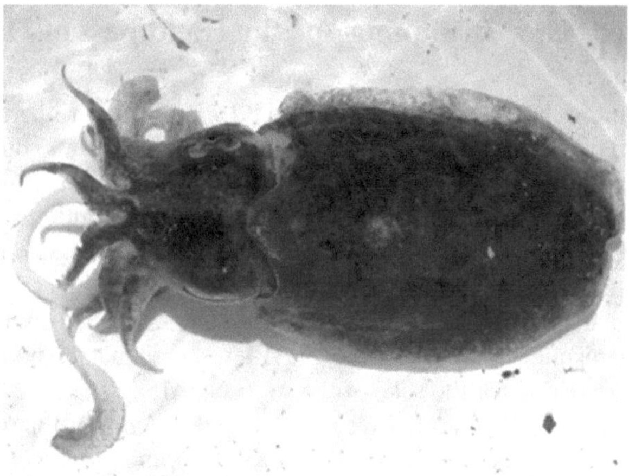

Fig. 6.19 A cuttlefish or sepia showing the legs arranged around the head. Aristotle claims that the cuttlefish has the "wickedest tricks and is the only one that uses the ink for concealment and not merely when frightened" (*HA* 621b29)

obviously aware of its use.[16] This seems to be a case of the abstraction from reproduction and sexual parts, even though the modified arm is not primarily a sexual organ.

[16]The specialized function of the hectocotylus arm of the octopus was one of those things discovered by Aristotle and then only re-discovered more than 2,000 years later.

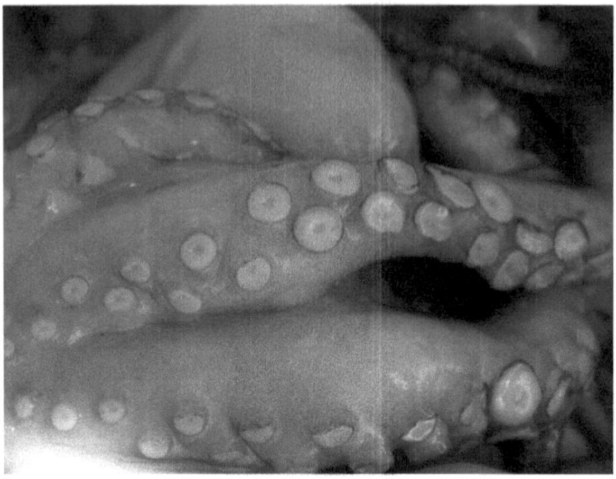

Fig. 6.20 The two rows of suckers of the common octopus (*Octopus vulgaris*). Aristotle distinguishes it from other smaller octopuses (*HA* 525a14)

6.8 Another "New Beginning" or the Handiness of Hands

The new beginning that is announced in *PA* IV.10 is a return to parts which have been mentioned but not described (685b30). It seems to be a turn to the human—we should note that the new beginning in Book II is also a turn to the human. We are told that humans are the only one of the animals to stand upright due to their being of divine nature and thinghood (*dia to tên phusin autou kan tên ousian einai theian*) (686a27). While we have had many statements regarding the function of particular parts of particular organisms, we have had very few explicit formulations of the *ergon* of an organismic whole. Of course, in *PA* I we heard that the body is an instrument (*organon*) existing for the sake of the soul (645b15–20); in the context of the new beginning of *PA* IV, Aristotle says that the work (*ergon*) of that which is most divine is to think and to be prudent (*ergon de tou theiotatou to noein kai phronein*) (686a29). While Aristotle does not use the word for cause (*aitian*), it seems that one could point to this passage as suggesting that the cause of this upright stature in humans is the divine.[17] However, we should note that Aristotle does go on to suggest a cause for such things which ties into several themes that have been directing the argument for some time, such as the continuum between plant and animal, and the relationship between the inorganic and the organic in terms of the elements.

The activities of thinking and of the general sense (*koinên aisthêsin*)[18] would move-with-difficulty (*duskinêton*) if there were a lot of body in the upper extremities

[17]This would be to understand the preposition "*dia*" with the accusative as indicating cause.

[18]"General sense" replaces *phronesis* in the new formulation.

of humans (686a30)[19]; the motion of humans, the motion of thinking, would be impaired if the arrangement of parts were not consistent with our upright posture. Being upright entails that the greater weight of the body be distributed to the lower extremities. When an animal is "top-heavy", so to speak, the body lurches forward to the ground of necessity, requiring front limbs for support (686a32). The relative proportion of weight between the upper and lower extremities, or rather, between the head and the vent, determines the difference between being two-footed or four-footed.

Aristotle uses the term "dwarfish" (*nanôdê*) to describe the animals which, when compared with humans, have the weight distributed more toward the "top" (686b4). "Dwarfishness" does not indicate diminutive size, but the proportion of body weight between upper and lower extremities. Interestingly, humans, in infancy, are dwarfish (686b11); as they age, humans grow through this "dwarfishness" characteristic of most other animals, becoming upright, and thus have the weight of the body distributed more to the lower extremities.[20]

The process by which the upright stature of humans is attained also helps to explain the presence of buttocks and fleshy legs. Nature is forced to distribute much of the corporeal stuff to the lower portion of humans so that the buttocks are formed and have the character they have (689b11). At the same time, the buttocks are useful for resting the body, which is something required for the bipedal humans: "Quadrupeds find it no trouble to remain standing, and do not get tired if they remain continually on their feet—the time is as good as spent lying down, because they have four supports underneath them. But human beings cannot remain standing upright continually with ease; the body needs rest; it must be seated" (689b19). Instead of buttocks, quadrupeds have tails, which have a variety of functions associated with them (690a).

The dwarfishness of the body mirrors dwarfishness of the soul, so that all other animals are less intelligent than human beings (686b23). The cause (*aition*) of this is that "in many of them the principle of the soul (*ê tês psuchês archê*) is difficult-to-move (*dukinêtos*) and bodily (*sômatôdês*)" (686b28). The examination of dwarfishness has illuminated the connection between the body and soul. The implication appears to be that human beings could not be thinking beings without the bodily form they have. Is Aristotle suggesting that the activity of thinking is the cause of the upright stature of humans?

The search for the cause of the standing of humans leads us to one of the elements, fire or heat. Heat is responsible for raising the animal body, so that animals with little heat will become many footed as the body falls forward in development (686b29). The end point (*telos*) when going from animals with relatively high heat (humans) to animals with less and less heat is a head without-motion (*akinêton*) and without-sensation (*anaisthêton*), which is on the way to becoming a plant (686b34).

---

[19]Isn't this the same principle which Socrates expresses in the basket in the *Clouds*?

[20]Isn't this the riddle of the Sphinx, man as the animal which moves on four legs, two legs and then three legs?

So far, the argument has moved from the divine and thinking nature of humans, to the arrangement and distribution of weight between upper and lower, to the discussion which puts the emphasis on heat, which in turn explains the continuum from animal to plant. This way in which to explain the animal-plant continuum moves from the high to the low and by means of an appeal to the elemental heat.

The appeal to heat explains several complexities: the cause of animals having two, many or no feet at all; why some living things are plants and some animals; and why human beings are the only one of the animals to stand upright (*orthon*) (687a2–6). The heat that causes man to be upright also allows for the hands to be free (687a7), since they are not required to support the weight of the body. Though the trunk of the elephant can function like a hand (e.g. 692b17), humans actually have hands. To the observer there is obviously something special about the possession of this part and the functions of which it is capable. This is perhaps why Anaxagoras suggests that it is on account of the hands that humans are the most prudent (*phronimôtaton*) of the animals (687a7). Anaxagoras makes the claim that the presence of the hands are causally related to human prudence; Aristotle, however, wishes to reverse this structure, claiming that the cause of humans having hands is to be found in prudence (687a9). While Anaxagoras would suggest that human prudence or intelligence is caused by the possession of hands, Aristotle has already given some indication in this context that the cause of human prudence has something to do with the heat causing our upright stature, which also causes the hands to be free.

That the unique character of the human hand has something to do with human intelligence is recognized by Anaxagoras; related to this is the way in which the hands point to the arts, to human making. The hand is an organ (*organon*) (687a10) that gives humans the capacity to acquire many arts through its great usefulness (687a21). The handiness of this organ (*organon*) makes it possible for humans to make other tools (*organoi*) in the pursuit of finding comfort and survival. The arts allow for the mitigation of certain "defects" that people point to in the structure of human parts; people claim, not rightly (*ouk orthôs*) according to Aristotle, that the structure of humans is not beautiful (*ou kalôs*) because humans are barefoot, unclothed, and without any weapon of force (687a25). The arts, which are related to the structure of the hand, allow one to respond to these "deficiencies" of human parts. In addition to having an opposable thumb (687b13), the capacity of the hand lies, in part, in the ability to make the many branches and divisions come together to form a one, a solid piece (678b8). The deficiency in the parts, for example, nakedness, combined with the peculiar and extraordinary character of the human hand, makes it possible and, in a sense, necessary for the emergence of the arts. Some lack combined with some extraordinary capacity helps explain the emergence of the arts.

The fact that the human requires only two limbs to stand accounts for the fact that the hands can remain free. The standing of humans also dictates that the feet have a certain character. Humans have the largest foot in proportion to the other parts of the body (690a28). The foot is divided into toes; this is good as a protection against injury to the whole foot.[21] The length of the toes when compared to the length of the foot stands in an inverse relationship to the length of the fingers compared to the length of the hand (690a30). In other words, while the foot is relatively large compared with the hand, the toes are small compared with the fingers. This relationship is according to *logos* (690a32) because "the *ergon* of the hands is to take hold and keep hold of things, and this is done by means of that part of the hands which bends, therefore the fingers must be long. The *ergon* of the feet is to get a firm and reliable footing; and to secure this the undivided part of the foot must be greater than the toes" (690a34). Giving a *logos* seems to include seeing how the structure of the part relates to the *ergon*.

6.9 A Move Toward Reproduction and the Parts Associated with Reproduction

Perhaps *PA* IV.10 is a new beginning because the description of these parts mentioned but not described includes an attempt to incorporate issues regarding reproduction. Semen, menstrual fluids, and urine are things that are fluid in nature (*hugra de tên phusin*), so they are discharged through the same parts. On the basis of their being fluid, the residual parts of semen, menstrual fluid and urine are discussed as the same and associated with the same parts. The radically different functions of these liquid residues is suppressed in this context—that is, the otherness in the role of these materials is not stressed, only that they are similar in being fluid residues. Even sexual generation and the parts involved are cast in terms of the elemental. *Genesis*, in perhaps its most important form, sexual generation, keeps intruding into the argument, in part, because of its striking absence. The argument of the *PA* seems to require putting off a serious investigation of the parts associated with reproduction and generation. A possible explanation for this need could lie in the relationship between the parts for reproduction and the issue of purpose: the parts associated with reproduction seem to exist, not so much for the possessor but for some other, for the offspring. The mammae exist for the sake of providing nutrition to some other, to the offspring (688a20).[22] Such parts point beyond the organism that possesses them. In other words, the wholeness of the animal is shown to be less than a complete, autonomous whole by an examination of the function of the parts for reproduction.

[21]The idea that a whole divided into parts might serve as a kind of protection against the infection or damage to the whole is also stated as applying to the kidneys (671b5 ff.).

[22]Even here the discussion of the mammae is a bit skewed insofar as the nourishment of the offspring is said to be an additional function (*heteron ergon*), as opposed to the primary function. Of course, perhaps it is just an additional function because many male animals have mammae and they are obviously not used for the nourishment of off-spring (see 688b30 ff.).

However, not only the parts and residues associated with reproduction point to this; the intake of *trophê* (which is also part of the *genesis* of animals) and its output show that things pass into and out of the whole organism. In other words, the animal is not a hermetically sealed whole. The fact that some of the parts associated with reproduction also function in the discharge of other residues forces us to consider both examples of the *genesis* of animals, the process by which food is incorporated into the body and reproduction. In both cases, we are shown the limits of the animal whole.

In males, semen and urine are discharged from the same part. However, discharge of semen requires an erect penis, while urine does not. In fact, a penis that was always extended would be a nuisance to the other parts. Copulation and urinating require two different "states" of the same part. It is in this context that Aristotle suggests that "it is not unclear that the shape of the parts (*ta schemata tôn moriôn*) is from the necessity of the function (*pros tên ergasian angkaiôs*) they perform" (689a20). Female quadrupeds do not have the same tension between the requirements of the shape of the part in urination and copulation that exists in male animals. Instead, female quadrupeds are retromingent, they urinate backwards, which is useful in copulation (689a32).

6.10 Some Parts of the Egg-Bearing; a Sign of the Importance of Reproduction?

The new beginning announced in *PA* IV.10 looks as if it is a turn to the human, but also perhaps a turn in the argument toward a form of genesis that has been suppressed to a certain extent, toward the parts and processes of animal reproduction. *PA* IV.11 begins by suggesting that we move to a discussion of the parts associated with the egg-bearing (*ôotokon*), which are blooded animals. The distinction in kind or *genê* between live-bearing (*zôotokôn*) and the egg-bearing is grounded in generation. While most egg-bearing blooded animals have four feet, there is one *genos*, the snakes, with no feet at all. Aristotle claims that the cause of the footlessness of the snakes is treated in the treatise on the *Progression of Animals* (*peri tês poreias tôn zôôn*). This suggests that the *Progression* is similar to the *PA*, in being an examination of causes. Yet why would these treatises be divided? Is there something about the theme of progression that requires special treatment? It seems as if snakes are unique—they make up a single *genos* (690b14)—in being footless. However, Aristotle tells us another way in which snakes are peculiar:

> A distinctive feature present in the snakes as opposed to kindred animals, is their ability to turn their head to the rear while the rest of the body is at rest. This is because (*aitian*), like insects, they are capable of coiling, so that their vertebrae are flexible and cartilaginous. Thus while they do this of necessity, owing to this cause, nevertheless it is also for the better (*tou de beltionons heneken*), i.e. for the sake of guarding against dangers from behind; for being long and without fee, they are naturally unsuited both for turning around and for watching for dangers from behind; for it is of no use to be able to raise the head, yet be unable to turn it. (692a3–7).

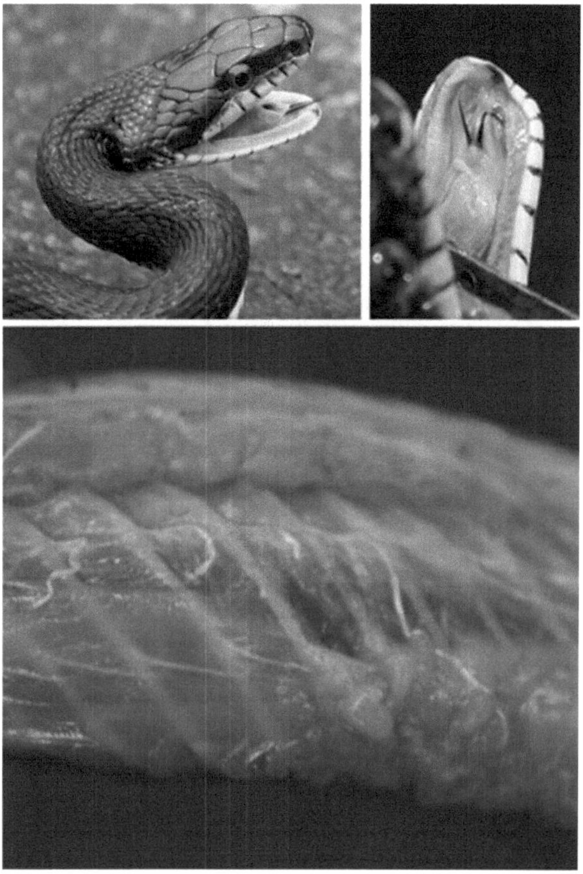

Fig. 6.21 Snakes are remarkably flexible. Aristotle suggests that the horny scales of snakes are made out the same material as bone and can become harder than bone (*PA* 691a19). The snake has a forked tongue which, Aristotle says, is peculiar and the tips of which are as fine as hairs (*HA* 508a28; *top right*). The flexibility of the system of bones is combined with the hardness of the covering. The skin and flesh have been removed (*bottom*) to show the system of ribs and vertebrae that allow remarkable flexibility (see Fig. 4.7)

Being footless causes several problems which are mitigated by the unique flexibility of the snake (Fig. 6.21).

This discussion of the parts of the egg-bearing animals concentrates on the head and the parts located in the head, such as the tongue, teeth, ears and eyes. The first thing Aristotle notes is the absence of a tongue in the river crocodile (690b20). The crocodile is both a land animal a water animal. This is the cause for its tonguelessness, since animals that live in the water do not have time to savor their food because of the watery environment (662a7). This leads to a discussion of the tonguelessness of fishes and the pleasantness of food. The cause (*aitian*) of the absence of a tongue in fishes relates to the fact that they cannot chew and enjoy

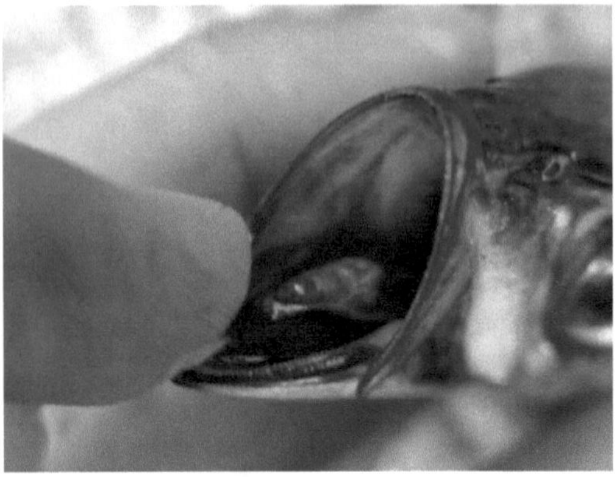

Fig. 6.22 This mackerel (*Scomber* sp.) clearly has something that looks like a tongue; although it is indistinct, it still functions as an organ of taste (*HA* 533a26). In speaking of fish tongues in general, Aristotle says "the tongue of fishes is prickly and not properly separate, while some fishes exhibit in that position a quite smooth and unarticulated surface unless you pull their mouth well open" (*HA* 503a2; cf. 505a30)

the pleasures of food (690b25). Aristotle does recognize that fish have tongues (e.g. 690b25, 662a7) (Fig. 6.22), but the key is that they cannot spend time chewing and savoring their food (662a7 ff.) so that perhaps the tongue is not allowed to serve in this function. Because fish tongues do not perform the function of tongues in discerning pleasurable food, Aristotle can say that fish do not have tongues. If we can say that a bird has ears because it has something which functions in hearing without actually having the part (see below the discussion of the "ears" in egg-bearing tetrapods), we can say that the fish does not have a tongue, even though we can see it in the mouth, because it does not function in taste. However, this does not mean that fish do not perceive the pleasantness of food. In fact, it seems as if all blooded animals have this capacity (690b35). There are two ways in which food and drink can be pleasant. There is the experience of pleasure when the food and oiliness of the food passes down the throat (690b30). There is also the experience of the pleasure of food with the tongue while chewing (690b25). All blooded animals have at least one of these capacities (for example, fish), while most have the power of both. The fact that there are two powers of discriminating the pleasant has the consequence that there are two kinds of intemperance (*akrasia*): the intemperance for food is not found in conjunction with the intemperance for drink and juices (691a2); the two means by which pleasure is discerned produce two forms of *akrasia*.

In egg-bearing tetrapods, the parts around the head, including the sensitives (*ta aisthêtêria*) are similar to those parts found in other animals. Their teeth are like the teeth of fish in that they are sharp (691a10). Their ears, although they function in hearing, do not look like ears, but like ducts, as in birds (691a11). They are ducts and not what we associate normally with the look of ears because of the hardness of their covering (*dermatos*) (Fig. 6.23). In birds and these egg-

Fig. 6.23 A giant tortoise of Lesvos. The tortoise is an animal that drinks little, yet still has a bladder which is surprising to Aristotle (676a30). The tortoise—one of the four-footed, egg-laying group—are "all covered with horny scales which correspond in position to the scales of fishes, but are harder in nature" (691a15). The scaly covering is clearly visible on the face and legs

bearing tetrapods the character of the covering affects what the "ears" look like, although they function in hearing. Thus, function, rather than look, determines what we call the ear. The material character of a certain part (the *dermatos*) determines the character and shape of another part (the hearing ducts or "ears" in birds and egg-bearing quadrupeds).

The eyes, eyelids, and sight of egg-bearing quadrupeds are compared to the birds. These animals do not have any upper eye-lids, which is similar to the case of birds (691a20). However, egg-bearing quadrupeds and birds differ markedly in the degree of their vision, which is a result of the relative hardness of their eyes (691a24). Keen sight is very useful (*chrêsimôtera*) for birds in their habits of life, as opposed to the egg-bearing quads that live in holes (*trôgloduta*) (691a25). The cause (*aitian*) of the better sight in birds is understood in terms of the usefulness to their habit of life.

The head in general has two divisions (*duo diêrêmenês*), the upper and lower parts of the jaw (691a27). Like fish and birds, the movement of the jaw in egg-bearing tetrapods is only up and down (691a29); this is in contrast to the movement of the lower jaw in humans and live-bearing quadrupeds which, in addition to moving up and down, moves from side to side (691a32). Both motions, up/down and side-to-side, can be useful depending on the teeth. The cause (*aitian*) of the egg-bearing tetrapods only having the up and down motion of the jaw lies in the fact that they possess only sharp teeth utilized in biting and tearing. They do not possess the grinding-like teeth, which would be utilized in the side to side motion possessed by humans and live-bearing quadrupeds. The side-to-side motion would be of no use to the egg-bearing tetrapods because they lack the corresponding parts, the grinder teeth. Having the side-to-side motion without the flat, broad teeth would be superfluous (*periergon*), and nature never makes anything that is superfluous.

The motion of the jaw in most animals originates in the lower jaw; the crocodile is an exception in that it moves its upper jaw so as to be able to compensate for the lack of serviceable hands (691b6). This arrangement in crocodiles allows it to seize and hold the prey on the one hand and eat it on the other; this is also given as the causes for the movements of the crab claw (691b14), although we should note that the crab has two parts, the claw and the mouth, to accomplish the two activities of seizing and eating. The crab, because it does not necessarily consume its food in the water, can have parts that divide up the labor of seizing and eating food; because of the constraints of the watery life of the crocodile, the two functions (seizing and eating) are packed into one part.

6.11 The Chameleon's Colors

We suggested that the octopus's evacuation of the liquid residue in fear highlighted the connection between soul and body (678b36). Because he has little blood, the chameleon is an egg-bearing tetrapod with very little flesh on his bones (692a20).[23] As Aristotle explains, the concoction of blood is responsible for fat and flesh and other more solid bodily substances. As a result of the process of concoction of the blood, the chameleon, with little blood, would have little flesh. However, having little blood is also the cause of the animal's habit of soul (*tou tês psuchês êthous esti tou zôou*) and is responsible for the polymorphism of the chameleon (692a21). Little blood leads not only to leanness, but also to coldness (692a24). Fear and the animal's response to fear are understood in terms of blood and heat. This is the way of the physiologist (cf. *De Anima* account of the boiling of the blood). The cause of fear (or habit of soul) and the cause of leanness are given in terms of the physiology of blood.

6.12 The Parts and Habits of Birds: The Movement from *kata tous bious* to *dia ton bion*

The description of the parts of the egg-bearing tetrapods often makes reference to the similarities and differences compared to birds. So, it is not surprising that the argument turns to a discussion of birds (*PA* IV.12). This discussion will not focus on the sense organs, since they have already been addressed (692b18). Instead, we are again confronted with the question as to why the discussion of parts needs to be segregated into sensitive and then non-sensititves (see 647a3 and the discussion of

[23]Ogle suggests that this is one of the few foreign creatures that Aristotle studied alive.

the distinction between instrumental (*organikôn*) and sensitives). The structure of much of the book then would hinge on this initial analysis, initial division between the instrumental and sensitive.

The differences (*diaphora*) in the parts of birds can be seen in terms of excess and defect (*huperochê kai elleipsei*) and according to the more and the less (*kata to mallon kai êtton*) (692b3). The notion of the parts of organisms within the same kind differing by the more and the less was introduced as a useful tool for those interested in dividing into kinds (644b9); the more and the less was also contrasted with the notion of analogy—for example, the way in which fish scales "resemble" feathers. It seems as if the more and the less points to the figure (*schêma*) and the look of the parts (644b9–12)—to resemblance in the strict sense—without reference to the living whole. Analogy, in contrast, points to the function of the parts, for instance fish scales and feathers offer a protective covering to the different looking organisms that possess them. When the parts are examined as part of the living of the organism, one can discover analogies more easily. In other words, the more and the less seem to point to looks, while analogy points to motion, the motion of function of the living. However, the emphasis on the look of the parts of animals when examining the more and the less might be, in a sense, put into motion.

The contrast between differences based on the more and the less and those based on analogy is an issue when Aristotle turns to birds. As has already been noted, the parts of different kinds of birds differ in excess and defect and these differences are according to the more and the less (692b3). However, when compared to other animals, the parts of birds are said by Aristotle to differ by shape (*tê morphê*) (692b8). One expects that the contrast with other animals will be explained in terms of analogy. Why isn't the contrast here between the more and the less and analogy as in the beginning of the book (644a12 ff.)? How is *morphê,* insofar as it means shape, a suitable replacement for analogy, which seems to be always pointing to function? Isn't *morphê* similar to the figure (*schêma*) that those interested in classifying animals appeal to (644b9)? *Morphê* might have a wider meaning than "shape" if we are right in contrasting the more and the less and analogy. Or perhaps we can resolve this apparent problem by noting that the issue is the otherness between birds and other animals and that analogy points to a certain sameness between different kinds of organisms.

The peculiarity (*idion*) of the shape (*morphê*) of birds involves their possession of feathers (692b10). Feathers, previously spoken of as analogous to the parts of other animals (644a24), now are peculiar or idiosyncratic. Another feature of the birds' peculiarity, their otherness, is the possession of the beak. The beak is a replacement, or a fusion, of two parts, in that it serves as teeth and lips (692b15), just as the elephant's trunk takes the place of hands and the tongue of certain insects replaces the mouth (692b17–19).[24] The beak is peculiar to the bird, it is part of its otherness;

[24]The word for "trunk" is *proboskis,* which is also used for the tongue-like appendage of the fly and the tentacles of cephalopods (Lennox 2001, p. 236).

Fig. 6.24 Avocets (*Recurvirostra* spp.) and flamingos (*Phoenicopterus ruber*) feeding in ponds near the Bay of Kalloni. The relatively long neck and correspondingly long legs of these birds are the typical proportions according to Aristotle Avocets have an upcurved bill, more evident in the picture of the American avocet (*top left*), that is used in feeding along the shore. It is somewhat peculiar to me that the there are very references to the flamingo in classical writings (see Thompson 1936, p. 304)

however, in comparing it to the trunk of the elephant and tongue of insects, Aristotle points, without making it explicit in this context, to the analogous ways in which such parts function. When just looking at dead specimens of birds, elephants, and insects, one would have no problem grouping them and distinguishing kinds, no problem ordering them. However, when one examines the peculiar movements and functions of various parts, that order is put aside for another inquiry that involves seeing how what one thought is other is, in some sense, the same.

In looking at the more and less of birds as a group, we expect differences of magnitude to be an issue. It seems as if the lengths of the neck and of the legs correspond, so that a bird with long legs will have a long neck (692b22) (Figs. 5.5 and 6.24). What use would long legs be to a bird with a short neck? The exception to this general relationship lies with web-footed birds (693a5). These web-footed birds have long necks for getting food out of the water and short feet that allow them to swim (Figs. 5.6 and 6.25). Their watery environment—eating from the water and swimming—determines that these web-footed birds do not abide by the neck-to-leg

Fig. 6.25 A number of European shags or cormorants (*Phalacrocorax aristotelis*). The long necks of these birds are not accompanied by correspondingly long legs. This is an exception to the general rule which Aristotle attributes to their peculiar life history. Cormorants swim low in the water and are very good divers, hunting fish. Aristotle notes that like a stork in size "except that its legs are shorter; it is web-footed and good at swimming, and black in color; it sits in the trees and nests there, the only such bird to do so" (*HA* 593b18–20). It is interesting that water birds take to trees to nest

ratio.[25] Habit of life dictates that these kinds are idiosyncratic in this regard. You have a general relationship (neck-to-leg length) that corresponds to life history and habit. However, there also exists a particular life history strategy in which it is useful to have a long neck and short legs. Thus, the neck-to-leg relationship can vary according to habit of life. The more general neck-to-leg length correlation does not hold here *because* of the watery habitat of these web-footed birds.

Likewise, beaks, which are another peculiarity of birds (692b15), can vary or be differentiated (*diaphoran*) according to life (*kata tous bious*) (693a11).[26] Straight beaks are useful for simple feeding; curved beaks are useful for the meat eaters who necessarily have to get their food from animals; broad beaks are useful for marsh

---

[25]It seems as if there are always interesting cases of animals in the muck between land and water, e.g. crocodiles, elephants, egg-bearing tetrapods, and these birds. The in-betweenness of the habitat produces unique and interesting kinds.

[26]I wonder if there is a difference between parts varying according to life (*kata tous bious*) versus variation according to logos (*kata tous logos*). The formulation "according to logos" seems to better support the notion of some kind of maker, a Nature that makes, whereas the other formulation puts the emphasis of life. Do these two point to different causal accounts?

Fig. 6.26 Black-winged stilt (*Himantopus himantopus*) with a young chick

birds for digging and pulling up food. The kind of beak also seems to relate to the neck-to-leg relationship—for example, curved beaks belong to, and are useful for, raptors with short legs and necks (Fig. 5.9). Perhaps the best example involves wading birds with long beaks and long necks used in catching small water animals[27]; the neck acts like a fishing rod while the beak acts like the line and hook (693a20) (Figs. 5.5 and 6.24). Usually we are presented with a picture of the explanation of some aspect of animals being understood in terms of some *technê*. In this case, the art of fishing is understood in terms of an understanding of the habit of these birds; perhaps an analysis of the organisms can improve the development of the arts, or perhaps watching a wading bird fish might be the origin of the art of fishing.

Perhapsthemost peculiar thing about birds is their mode of locomotion. Birds possess wings, which allow them to fly. Also, like human beings, they are bipedal (692b24). However, while humans and birds are both bipedal, the two groups differ in the way their legs bend; legs bend inwards (*eisô*) in birds, outward (*exô*) in human beings (692b4) (Fig. 6.26).

We can also note that, while birds may be bi-pedal, they are dwarf-like and do not really stand upright (695a5). Because they are blooded creatures and four is the greatest number of motion points (*sêmeiois kinêsontai*) for blooded creatures, they possess two legs in addition to their two wings (692b10).[28] Flight-ability (*ptêtikon*)

[27] Again, the organisms that straddle the wet and the dry seem to offer illuminating examples in the causal analysis of the parts of animals.

[28] Are the hands and arms of humans considered "motion points"? This would make the discussion of hands in terms of thinking and the arts (687a7 ff.) very interesting.

Fig. 6.27 Talons and beak of buzzard (*Buteo buteo*) (*top* and *middle*) and a short-toed eagle (*Circaetus gallicus*) hovering in search of snakes and lizards (*bottom*)

is in the thinghood (*en tê ousia*) of bird. Thus, out of necessity (*ex angkês*), they have two wings in order to be stretchable (*tonikoi*) (692b10–15).[29]

There exist certain differences within the tribe of birds with regard to flying. Some birds are strong fliers, while others are not. The good fliers have big and strong wings (694a); they are also the birds with talons, crooked beaks and are meat

[29]In his note in the Loeb, Peck says "The chief difficulty in translating this passage is due to the word *tonikoi*, a jargon-adjective in –ikos, which seems to have been suggested to Aristotle's mind by the similar adjective *ptêtikon* in the next line."

Fig. 6.28 Great white pelican (*Pelicanus onocratulus*) on the Gulf of Gera. This bird is such a poor flyer that it takes several big hops, utilizing the large webbed-feet, on the surface of the water before it is fully airborn. This bird, like the European shag, offers another example of the exception of short legs and relatively long necks

eaters. It is necessary (*anagkê*) that they are good fliers because of their lives (*dia ton bion*)[30] and for-the-sake-of (*heneka*) this they have many feathers and big wings (694a3–5). The particular life-style of these birds causes or makes it necessary for them to be strong fliers which necessitates that they have those parts which enable strong flying. I would suggest that the account given here replaces the maker, the Nature that never makes anything in vain, with a causal account that is situated in life. The meat-eating way of a raptor, a bird of prey, is understood to be cause of the arrangement of parts, such as the crooked beak, the talons, the feathers and wings necessary for swift flight (Figs. 5.8 and 6.27).

Other kinds (*genê*) of birds that are good fliers include those whose safety relies on the ability to escape through flight (694a5). The predators and prey both have the ability to fly quickly and strongly. Is Aristotle suggesting that way of life is a cause in some sense? Perhaps the best way to understand these observations about birds is to suggest that Aristotle recognizes that parts and habits are maintained because of the life-style of the organisms; that is, life-style is the cause of the maintenance of parts across generations.

On the other side, we have poor fliers, which are usually heavy birds (Fig. 6.28). Their life is spent on the ground eating fruit or getting-food (*bioteuma*) around water. As a means of defence, some poor fliers have spurs; these spurs are useful in fights on the ground (694a16). A bird has either spurs or talons, never both. In fact, talons on a heavy bird are not only useless but would be harmful (*blaberos*), since the claws would stick into the ground and impede progression (694a17). The fact that birds never have both talons and spurs is part of the necessity of their *genesis* (694a22), where *genesis* refers to the actions of the material. There is earthy (*geôdes*) matter in their bodies that courses and generates (*ginetai*) parts useful for weapons. Where this earthy matter courses upward it produces a large or hard beak; when this earthy matter courses downwards it produces spurs or talons, long legs or

[30]The preposition "*dia*" with the accusative tense suggests a causal relationship.

webbed feet (694a24–b6). Aristotle gives an account of the presence of beaks, spurs and talons in terms of the movement of the material; the material is not moved into place and shaped by a maker, but rather has a movement of its own. But actually, the account explains why a bird would have a beak, but it does not explain why a bird has the beak it does, why it is long and narrow or broad or curved, etc. In other words, the account that relies on this movement, this coursing of the earthy material does not explain the shape of the part, only its presence. What causes form? Do we have to appeal to a maker-Nature at this stage?

The account of the presence or absence of things like spurs is given in terms of necessity but can also be couched in terms of what is good. Webbed feet are caused by necessity but also on account of the best (*dia to beltion*) and are for-the-sake-of (*charin*) their life, living in water as they do (694b7) (Figs. 6.25 and 6.28).

If the necessary coursing and movement of the material accounts for the presence of a part, does the good or the best give us a cause of the shape and form of that part? If so, how does this work? Let me suggest that we get one answer to this question when Aristotle suggests that some birds have long legs and the cause (*aition*) of this is that the life of these birds is marshy (694b12). The way in which an organism lives must point to the best; but in this context we have a clear statement suggesting that it is the way that causes a part to be the shape and form it takes. In the case of birds, the coursing of material can explain the presence of a part, but the way of life explains the shape of the part.

Having given us the means by which to understand how the necessary and life-style, which entails what is best, can explain the arrangement of parts in birds, Aristotle seems to obscure this account by falling back on the Nature that makes (*ê phusis poiei*): "Nature makes the organs to suit the work (*ta organa pros to ergon*), but not the work to suit the organs (*ou to ergon pros ta organa*)" (694b14). In the first case, we have the cause being related to the life, the marshy life. In the second case we have nature making organs for work. In this context all of the organs of the body are made for work, which has to be interpreted as life. The argument suggests that life and life style can replace the Nature that makes. Nature is said to make *ta organa* and not *to ergon*. The question of nature as cause and life as cause can be seen to divide along which perspective one adopts, the instrument and organs or the *ergon*, which has to be work or life of the organism. Can Nature make the *ergon*? To put the question a slightly different way, is there some teaching about the relationship between nature and *ergon* in the infamous phrase "nature never makes anything in vain (*peri**ergon***)"? And what would it mean if nature could not make the *ergon*? One might be tempted to appeal to the making Nature (*ê phusis poiei*) when one is looking at the organs, the instruments (*ta organa*) because the very word reminds us of the making or the craftsman.

Aristotle immediately gives us an example to help in determining what is meant by this passage. Aristotle makes the suggestion about the making of the organs to suit the work (*ergon*) in his discussion of marsh birds. Wading birds stick out their legs when flying (694b21) (Fig. 6.29). They do this because they are missing tail feathers. Generally, the job of aiding in steering falls to the tail feathers, but in the case of wading birds, the legs accomplish this. Why, in this case, did nature not make

Fig. 6.29 An egret (*top left*), a white stork (*Ciconia ciconia*) and a heron all with legs extending behind short tail feathers. Aristotle notes that "among birds with long necks, those with a thicker one fly with the neck stretched out, while those with a thin, long neck fly with it bent up; for on account of this protective device the neck is less easily broken if they fly into something" (694a26). The former can be seen in the stork while the latter is true of the heron's position in flight

the organ for the work? The work, helping to steer in flight, needs to be done. The answer, in part, lies in the necessary constraints on nature; there is not a sufficient amount of nutriment (694b2) to make both tail feathers and long legs. The function here, the aid to steering and maneuverability, is accomplished by legs, but perhaps accidentally. It is perhaps a fortunate break that the legs are capable of taking on this extra function. We are given another case in which one part is pressed into service for a number of different functions. But can we say that Nature makes the organs for the function in this case? There appears to be work that is done by organs not designed for the specific job, e.g. legs as tail feathers.

6.13 The Genos of Fish

Aristotle turns from a discussion of birds to a discussion of the external parts (*tôn ektos moriôn*) of fish. He will take a moment, as we will see, to discuss those organisms that dualize, that partake of the water and the land. Fish do not have separate limbs attached to their bodies since they by nature go through the water—their nature is to travel through the water (*dia to neustikên einai tên phusin*) and is according to the logos of their thinghood (*kata ton tês ousias logon*)

Fig. 6.30 Conger eels (*Conger conger*) and *Lithognathus murmura*. Note the gills and the pelvic and pectoral fins on the latter, whereas the conger eels do not have any gills and they also lack the pelvic fins on the belly. The eels that Aristotle claims can move about outside of the water as if they were land creatures are not these conger eels, but common eels (*Anguilla anguilla*), as in Fig. 6.32

(695b17). Nature is used in this instance as something relevant and determinative of the beingness of fish. Aristotle continues from this to point out that "it is said (*eipei*) that Nature (*ê phusis*) never makes (*poiei*) anything superfluous (*periergon*) or not needed" (695b19). With this formulation and the proceeding one, we are confronted with two very different notions of nature, nature as something related to the substantial being of an organism, and Nature as maker. I wonder if the general formulation, "nature makes nothing in vain," is the general opinion regarding the world. This would be the formulation for the recognition of the order that appears manifest. The inquiry into nature moves from the notion of Nature as maker to the notion of nature that points to the *ousia* of the organism.

Fish have fins, and not legs, because they are swimmers. But we should note that Aristotle does recognize those sea creatures that use their legs to swim. Fish, generally, have four fins, because they are blooded creatures and have no more than four motion points (698a). The fins that Aristotle refers to come in pairs, two situated more or less on the underside (the pelvic fins) and two situated on the sides of the body (the pectorals) (Fig. 3.4); Aristotle does not consider here the median fins (the dorsal and anal fins) as part of the complement of motion points. There exist fish in which one or both of these pairs of fins is missing, such as the flat fishes and rays (e.g. *batos kai trugôn*) (695b30) and the serpent-like fishes, like the eel and the conger (*engchelus kai gongros*) (696a5; Fig. 6.30). We are not told exactly why the former, the flat fishes, lack the normal complement of fins, but Aristotle would presumably suggest that their lives on the bottom of the sea would make the pelvic fins a nuisance. With regard to the latter fishes, the serpent-like, the cause of their finless-ness is stated in another work, the treatise "on the progression and movement of animals."[31] Aristotle says that the cause of fish lacking fins is explained in the *Progression*, but then goes on to outline the cause in the context of this discussion:

[31] *De Incessu* 709b7, 708a9ff, and cf. *PA* 690b16.

Fig. 6.31 A bream (*Diplodos sargus*) with a terminal mouth, a mouth positioned at the end of its snout; a striped red mullet (*Mullus surmuletus; top right*) with a slightly sub-terminal mouth (see Tipton 2008 for a more detailed account of Aristotle's observations regarding its feeding); the upturned mouth of the European barracuda (*Sphyraena sphyraena bottom left*). A weever fish (*Trachinus draco, bottom right*) buried in the sand, its upturned mouth allowing for such a life on the bottom. The weever is an example of an upturned, gaping mouth, while the barracuda would be an example of what Aristotle has in mind in describing an upturned, razor-toothed mouth (697a)

the limit to the number of motion points combined with body shape dictate that they have no fins. Serpent-like fishes are long in the body and only four motion points would be too spread out. The extraordinary length in body, combined with the fact that they are essentially blooded creatures with no more than four motion points, requires that they move about in an undulating way (Fig. 6.30). We can note that the cause behind their having no fins is the same as the serpent's footless-ness (696a5) (Fig. 6.31). Serpent-like fishes crawl or slither and thus are able to move about on land (Fig. 6.32).

A peculiarity that all fish share lies in their having gills, which determines the *genos* (696b).[32] Whether a fish has many or few gills depends on whether much or little heat is in the heart: Having few gills (which implies little heat) allows the eel and similar fishes to live outside of water for a long period. The eel is almost a land animal in its ability to move about outside of the water and to stay out of the water for long periods of time. Again, we are presented with a very interesting organism,

[32]See *Respiration* 476a1, 480b13 for a statement of the causes. Again, why aren't the causes discussed here in the *PA*, which is the stated purpose of the treatise?

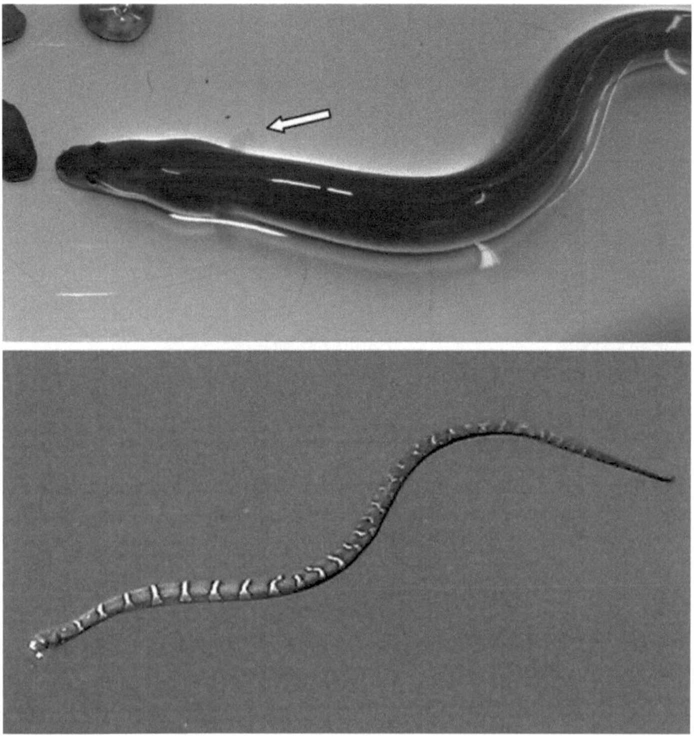

Fig. 6.32 A freshwater eel (*Anguilla rostrata*) undulates through a pan with very little water (*above*). The eel is the only fish named in Homer (Thompson 1947, p. 59) Note the small, relatively reduced pectoral fins set close to the barely visible gills (*arrow*; *HA* 489b27). The ability of the eel to move about on land made an obvious impression on Aristotle (696a5). Just in front of the reduced pectoral fins is the covering, the operculum, for the gills. The fact is that fishes such as the eel with few, less powerful gills are able to live out of the water for extended periods (696b20). With respect to their undulating motion, fishes like sea-eels have "absolutely no fin, but move by bending, using the water as snakes use the land; indeed snakes swim the very way that they slither on land" (696a6). The movement of the water snake (*below*) illustrates the similarity of movement exhibited by the snake in both water and on land, but also the similarity between certain types of fish and snakes

one that is atypical, that straddles the border between two realms, in this case the watery and the dry. Aristotle concludes the discussion of the parts of animals with a discussion of several animals which straddle worlds: dolphins, seals (Fig. 6.33) and whales, which are in between land and sea; bats (Fig. 6.34), which are between land and air; likewise, ostriches, which are between land and air, although in a different manner. These in-between creatures have unique parts and arrangements of parts that allow them to flourish in the in-between.

Of course, fish, like all other animals, require the parts associated with acquiring food and nutriment. There are differences (*diaphora*) with regard to the shape of the mouth of fishes (696b24). Some have mouths situated at the end of their snouts,

Fig. 6.33 The flippers of a seal showing their peculiar character. Insofar as they are in between land and water, "they partake of both and of neither" (*PA* 697b4)

Fig. 6.34 Aristotle argues that "bats as flyers have feet, but as four-footed they do not; and they have neither tail nor rump—no tail owing to being a flyer, no rump owing to being a land-dweller. And this happens to them of necessity; for they are membranous-winged, and nothing has a rump unless split-feathered" (*PA* 697a6–8). While the hind feet are clear (*bottom right*), the membrane stretched between the five fingers makes the front feet almost unrecognizable (*top*). This extended, membranous "foot" is hardly a foot at all when compared to other four-footed animals. It is, as I suspect Aristotle would argue, closer to the webbed foot of a bipedal water bird

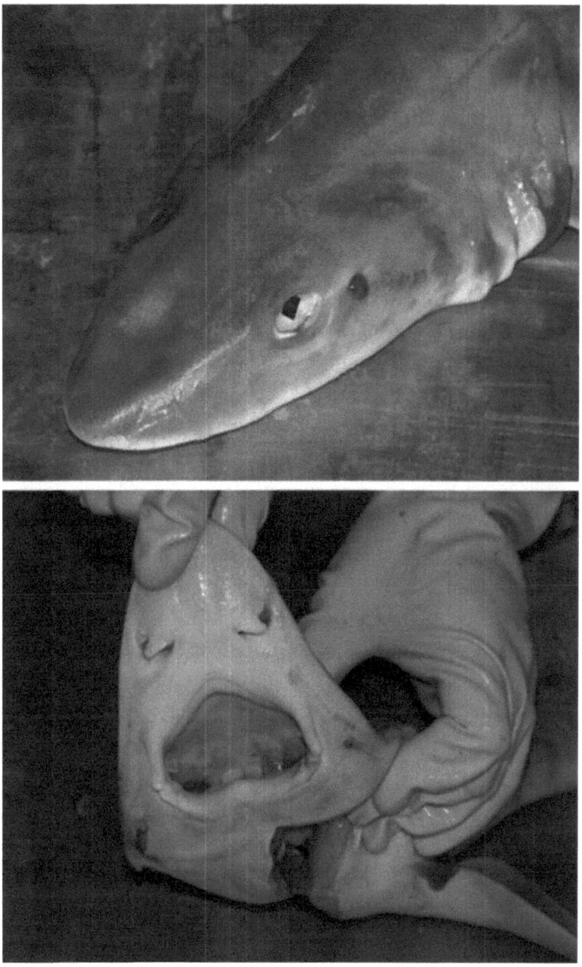

Fig. 6.35 A spiny dogfish (*Squallus acanthias*) in a Mytilini fish market. The sub-terminal mouth of this shark must have impressed Aristotle

some have mouths that are turned up, some situated below the snout (Fig. 6.31). It appears (*phainetai*) that Nature made (*ê phusis poiêsai*) carnivorous fishes with mouths below for the sake of preserving other animals, because while these fishes must turn over on their backs to feed, the prey has time to escape (696b27) (cf. Ogle's note 29, p. 251; Fig. 6.35). This is the only instance of a character of an animal being for the sake of another. This seems to be a strong case for the working of a general teleological scheme by which the parts of animals are constructed in light of the parts and habits of other animals' needs. But it is a matter of *phainetai*, an appearance. Could the cause of the arrangement of this type of mouth be what is good for another kind of organism? Reading on, we see that Aristotle gives additional suggestions as to the cause of the sub-terminal mouth. The mouth is placed on the underneath so as to prevent these animals from eating too much and being destroyed in the process (696b30). Additionally, the nature of the snout (*tên tou hrungchous phusin*) is such that it would not allow for the placement of the mouth at the end of the snout (696b33). The arrangement and character of the snout causes the mouth to be what it is. This is an example of one part necessarily dictating the arrangement and placement of another part; it should not be surprising that a whole of parts requires that the parts interact and have some causal effect on other parts. The whole determines the parts but also the parts within the whole determine the shape and character of other parts. The appearance of Nature as maker (696b27) is contrasted with the nature (*tên phusin*) of the snout. The nature of one part, such as the snout or the thickness of the octopus leg (685b15), constrains the arrangement of another part, such as the sub-terminal placement of the mouth in certain fishes or the octopus having only one row of suckers (685b15). This is the constraint of necessity, which is also the constraint of nature.

This constitutes a peak in the argument of the *PA*. One is presented with the appearances of a nature that has arranged it so a "defective" part in one organism is for the sake of other organisms. However, we learn that the "defective" part prevents the destruction through gluttony of the organism with the said part. It is further explained that the nature of the snout prevents any other placement of the mouth. We move from a making Nature, to an explanation in terms of what is best for the organism, to an explanation grounded in the necessary interaction of the parts of the organism within the whole. This movement of the analysis of animals is one that can be described as a philosophical zoology. One moves from the wonder of a making Nature to a more subtle, if incomplete, description of the causes of the parts of animals. In this movement, we learn as much about the way we think as we do about the organisms of study. When examining the mouths of sharks, our first inclination is to describe its placement in terms of the supposed benefit to its prey. The assumptions behind this explanation are similar to the one that suggests that plants exist to feed animals, and animals exist to feed and clothe humans (*Politics* 1256b16). The explanation of the shark's mouth that focuses on the benefit to prey is one that avoids finding fault with nature, but it is also the one that does not recognize what is best for the organism or the constraints of necessity. As the movement of this account shows, Aristotle's *PA* analyzes the causes behind the parts of animals at the same time as illuminating the cognitive lens through which we examine the organic world.

6.14 Concluding Remarks

As we have seen, a close analysis of the argument that Aristotle presents regarding the cause of the mouths of sharks and other pisciverous fish is very illuminating: the placement of the mouth is described first as being present for the sake of prey, then as a result of the need to prevent self-destructive gluttony, and finally as necessary because of other parts, in particular the snout. By carefully exploring the movement of the argument of this particular passage we see the way in which different causal accounts unfold as the attempt is made to understand the sub-terminal mouths of sharks. It is noteworthy that the final stage of the argument presents us with an example of the way in which parts of wholes can affect the characteristics of other parts.

But what have we gained by a careful analysis of the whole argument of the *PA*? We certainly learn more about the organic world and about the way in which we view that world. Our initial surprise at the level of detail of Aristotle's biological investigations—for example, his observations on the reproductive parts of echinoderms—gives way to an appreciation of the way in which ontological problems are drawn out of these musings. This is the foundation for the activity of philosophical biology.

Analysis of the *PA* has provided numerous examples, in addition to the shark's mouth, of one part causing the structure or arrangement or even the presence of another part. Because the process of concoction associated with the stomach and other viscera produces a relative abundance of heat, the diaphragm is structured in such a way as to divide off these from other more sensitive parts, especially from the source of the sensory soul. The activity of the whole requires a dynamic balance between parts that, on their own, might be injurious.

One example of a part, or rather the defectiveness of a part, necessitating another part is given in the case of the crop. Birds require crops for processing their food because of their deficient mouths. In a similar vein, camels have more than one stomach because of the nature of their food in addition to their deficient mouths. The nature of their food—a characteristic of the camel's life history—combined with its relatively deficient mouth makes necessary multiple stomachs. These examples—bird crops and camel stomachs—not only highlight the way in which one part has a causal effect on other parts, but they also illuminate the way in which the defective is understood as a cause. This defective cause is only a cause because the phenomena are wholes of parts, so it might be better to understand this as the whole causing the partness of the parts. In any event, what is traditionally understood as Aristotle's "doctrine of causes" looks oversimplified in light of the examples explored in the biological works.

There has also emerged, in the course of the argument of the *PA*, discussion of animal parts that can be understood as wholes themselves. The skeleton, with its bones and joints, is a whole that is embedded in a larger organic whole. In addition to its role in overcoming the divide between the homogenous and non-homogenous parts, the heart is also described as an articulated whole. In both cases, it is motion—the motion of the animal made possible by an articulated whole and the motion of the beating heart—that allows us to see these parts as wholes. The motion of animals and their parts, in contrast to the fixed motions of the heavenly bodies, suggests a complexity in which something that is viewed as a part on one level is understood as a kind of whole on another.

A related ontological issue that emerges from a careful examination of the *PA* is the question of *genesis* or becoming. *Genesis* is understood, in the *PA*, in terms of the continuous maintenance of the whole organism. By means of concoction, the blood allows for the incorporation of nutriment into the fiber of the animal. This maintenance, rather than animal reproduction, is the form of *genesis* that is of primary interest in the *PA*. Aristotle puts off a discussion of genesis as reproducing offspring for the *Generation of Animals* (GA). This gives us a perspective on the relation between the *PA* and the GA; each discusses different aspects of the *genesis* of the organic. In each case, *genesis* illustrates the way in which organic things—as self-reproducing and as reproducing offspring—are fundamentally different from the inorganic.

The diversity of ways in which animals feed and get their nourishment provides Aristotle with much to chew on, but also provides an opportunity to examine what the activity of *historia* entails and how the *PA* relates to the *History of Animals*. As we saw in his discussion of hard parts, Aristotle makes a distinction between the bones of carnivores and those of herbivores. The bones of a carnivore, Aristotle observes, are necessarily harder than the bones of an herbivore. This can be explained by a difference in the way each gets food. Aristotle attempts to understand an organism's osteological and morphological characters with respect to aspects of its body size, its habitat, its mode of locomotion, and other aspects of its life history; I use the term "life history" to point to the broader issue of *historia*. In asking these questions and making these observations, Aristotle offers a causal account based on the work (*ergon*) or life history of the organism: carnivores have harder bones *because* of the way in which they get their food. Similarly, in the discussion of bird parts, Aristotle presents a principle, of sorts, that recognizes that long legs are usually present with long necks and that short necks and short legs influence the body plan of a particular kind. This generates a general relationship (neck-to-leg length) that corresponds to life history and habit; for example, long legs accompanied with a long neck allows wading birds to fish for their food in marshy areas. However, in the case of certain water birds, in contrast to wading birds, there exists a particular life history strategy in which it is useful to have a long neck and short legs, which is to break the general relationship. Thus, the neck-to-leg relationship can vary according to habit of life. The more general neck-to-leg length correlation does not hold here *because* of the watery habitat of these web-footed birds. These two cases—the bones of carnivores and the neck-to-leg ratio—seem to be cases in which *historia* provides a causal explanation which is significant in thinking about the relationship between the *HA* and the *PA* with respect to determining cause. We might hypothesize that the cause of the life styles was a result of the characters the animal has, instead of the reverse relationship where habitat seems to cause structures of animals. Of course Aristotle recognizes the need for some structure to be physically present, but it is only when the organism is in its environment that a particular structure will be shown to be useful and thus maintained. Aristotle recognizes the way in which environment, the particular habitat an organism is embedded in, has a causal role in the organization of the parts and habits of that organism. The account that explains the presence of certain parts by the mechanical movement of matter, the coursing of the earthy material for example, does not explain the shape of the part, only its presence. The functional advantage a particularly shaped part provides to an animal in its environment is a powerful tool for analysis.

The movement of the argument of *PA* II allows us to examine the issue of the relation between the organic and the inorganic. The way in which *PA* II divides into two parts marked by the announcement of a new beginning (II.10) directed our attention to the substance of the argument. The failure of the attempt to understand the animate simply in terms of the inanimate is indicated by the absence of the organic whole and by the appeal to the divine that marks the new beginning. The argument of *PA* II presents the organic in light of both the sub-organic elements and the supra-organic divine. Each perspective illuminates the organic sphere that is infused with generation and corruption. We can broaden this conclusion to apply to the book as a whole: this study has demonstrated the way in which, through a careful interpretation of the *PA*, matters and themes of concern to the ontologist manifest themselves in the exploration of the zoological world.

Bibliography

Editions, Translations and Commentaries

Aristotle, On the Parts of Animals, On the Movement of Animals, On the Progression of Animals, History of Animals, Generation and Corruption, Metaphysics, Physics, De Anima, Nichomachean Ethics; Plato, Philebus, Sophist, Statesman, Timaeus; Xenophon, Memorabilia.

Apostle, Hippocrates, and Lloyd Gerson. 1991. *Aristotle: Selected works*, 3rd ed. Grinnell: The Peripatetic Press.

Balme, David. 1991. *Aristotle. History of animals, books VII–X*. Cambridge, MA: Harvard University Press.

Balme, David. 1992. Aristotle. De Partibus Animalium I and De Generatione Animalium I (with passages from II. 1–3). With a Report on Recent Work and an Additional Bibliography, ed. A. Gotthelf. Oxford: Oxford University Press.

Balme, David. 2002. *Aristotle: Historia animalium, Volume I, books I–X text* (Prepared for Publication by A. Gotthelf). Cambridge: Cambridge University Press.

Cooper, J. 1987. Hypothetical necessity and natural teleology. In *Philosophical issues in Aristotle's biology*, ed. Allan Gotthelf and James G. Lennox, pp 243–274

Lennox, James. 2001. *Aristotle: On the parts of animals I–IV*, Clarendon Aristotle series. Introduction and commentary (trans). Oxford: Oxford University Press.

Nussbaum, Martha C. 1978. *Aristotle's De Motu animalium*. Princeton: Princeton University Press.

Ogle, William. 1882/1987. *Aristotle on the parts of animals*. Introduction and notes (trans). New York/London: Garland,

Peck, A. L. 1961. *Aristotle: Parts of animals,* Loeb classical library. Introduction, text (trans). Cambridge, MA: Harvard University Press.

Peck, A. L. 1963. *Aristotle: Generation of animals,* Loeb classical library. Introduction, text (trans). Cambridge, MA: Harvard University Press.

Peck, A. L. 1965. *Aristotle: Historia animalium, books I–III,* Loeb classical library. Introduction, text (trans). Cambridge, MA: Harvard University Press.

Peck, A. L. 1970. *Aristotle: Historia animalium, books IV–VI,* Loeb classical library. Introduction, text (trans). Cambridge, MA: Harvard University Press.

Thompson, D'Arcy W. 1910. Historia animalium. In *The works of Aristotle*, eds. J. A. Smith and W. D. Ross, Vol. IV. Clarendon Press.

J.A. Tipton, *Philosophical Biology in Aristotle's Parts of Animals*, Studies in History and Philosophy of Science 26, DOI 10.1007/978-3-319-01421-0,

Books and Articles

Balme, David. 1970. Aristotle: natural history and zoology. In *Dictionary of scientific biography*, vol. 1, ed. Charles Gillispie, 258–266. New York: Charles Scribner's Sons.
Balme, David. 1987a. The place of biology in Aristotle's philosophy. In *Philosophical issues in Aristotle's biology*, ed. Allan Gotthelf and James G. Lennox, 9–20. New York: Cambridge University Press.
Balme, David. 1987b. Aristotle's use of division and differentiae. In *Philosophical issues in Aristotle's biology*, ed. Allan Gotthelf and James G. Lennox, 69–89. New York: Cambridge University Press.
Balme, David. 1987c. Teleology and necessity. In *Philosophical issues in Aristotle's biology*, ed. Allan Gotthelf and James G. Lennox, 275–286. New York: Cambridge University Press.
Balme, David. 1987d. Aristotle's biology was not essentialist. In *Philosophical issues in Aristotle's biology*, ed. Allan Gotthelf and James G. Lennox, 291–312. New York: Cambridge University Press.
Balme, David. 1992. *Aristotle De Partibus Animalium I and De Generatione Animalium I* (first published in 1972), Clarendon Aristotle series. New York: Oxford University Press.
Ben-Tuvia, A. 1985. The impact of the Lessepsian (Suez Canal) fish migration on the Eastern Mediterranean ecosystem. In *Mediterranean marine ecosystems*, ed. Maria Moraitou-Apostolopoulou and Vassili Kiortsis, 367–375. New York: Plenum Press.
Blondel, J., and J. Aronson. 1999. *Biology and wildlife of the Mediterranean region*. Oxford: Oxford University Press.
Bolton, Robert. 1987. Definition and scientific method in Aristotle's posterior analytics and generation of animals. In *Philosophical issues in Aristotle's biology*, ed. Allan Gotthelf and James G. Lennox, 120–166. New York: Cambridge University Press.
Bolton, Robert. 2010. Biology and metaphysics in Aristotle. In eds. J. G. Lennox and R. Bolton, pp. 31–55.
Broadie, S. 1990. Nature and craft in Aristotle's teleology. In eds. D. Devereux and P. Pellegrin, pp. 389–404.
Brock, R. 2004. Aristotle on sperm competition in birds. *Classical Quarterly* 54: 277–278.
Buddington, Randall K., and Jared M. Diamond. 1986. Aristotle revisited: The function of pyloric caeca in fish. *Proceedings of the National Academy of Sciences* 83(20): 8012.
Burger, Ronna. 1984. *The Phaedo: A Platonic Labyrinth*. New Haven: Yale University Press.
Chalmers, Alan. 2009. *The scientist's atom and the philosopher's stone – How science succeeded and philosophy failed to gain knowledge of atoms*. New York: Springer.
Code, A. 1986. Soul as efficient cause in Aristotle's embryology. *Philosophical Topics* 15: 51–60.
Cooper, J. 1985. Hypothetical necessity. In ed. A. Gotthelf, pp. 151–167.
Daly, Lloyd. 1961. *Aesop without morals*. New York: Thomas Yoseloff Press.
Darwin, Charles. (2003, originally published 1859). *On the origin of species: A facsimile of the first edition*. New York: Wildside Press.
Darwin, Charles. (2004 originally published in 1879). *The descent of man, and selection in relation to sex*. New York: Penguin Books.
Daston, L. 2000. *Biographies of scientific objects*. Chicago: University of Chicago Press.
Davidson, James N. 1997. *Courtesans & Fishcakes: The consuming passions of the classical Athens*. New York: St. Martin's Press.
Davis, Michael. 1999. *The autobiography of philosophy: Rousseau's 'the reveries of the solitary walker'*. Lanham: Rowman & Littlefield.
De Pirro, M., G.M. Marchetti, and G. Chelazzi. 1999. Foraging interactions among three benthic fish in a *Posidonia oceanica* reef lagoon along the Tyrrhenian Coast. *Journal of Fish Biology* 54: 1300–1309.
Devereux. D., and Pierre Pellegrin. eds. 1990. *Biologie, logique et métaphysique chez Aristotle*. Paris

Dudley, John. 2012. *Aristotle's concept of chance: Accidents, cause, necessity, and determinism*. Albany: SUNY.
Frede, Michael. 1980. The original notion of cause. In eds. M. Schofield, M. Burnyeat and J. Barnes, pp. 217–249.
Freeland, Cynthia A. 1987. Aristotle on bodies, matter, and potentiality. In *Philosophical issues in Aristotle's biology*, ed. Allan Gotthelf and James G. Lennox, 392–407. New York: Cambridge University Press.
Freeland, Cynthia A. 1991. Accidental causes and real explanations. In ed. L. Judson.
Freudenthal, Gad. 1995. *Aristotle's theory of material substance: Heat and pneuma, form and soul*. New York: Oxford University Press.
Frost, Frank J. 1968. Scyllias: Diving in antiquity. *Greece & Rome* 15(2): 180–185. 2nd Ser.
Furth, Montgomery. 1987. Aristotle's biological universe: an overview. In *Philosophical issues in Aristotle's biology*, ed. Allan Gotthelf and James G. Lennox, 21–52. New York: Cambridge University Press.
Ghiselin, Michael. 1974. A radical solution to the species problem. *Systematic Zoology* 23(4): 536–544.
Ghiselin, Michael. 1985. Can Aristotle be reconciled with Darwin? *Systematic Zoology* 34(4): 457–460.
Gill, Mary Louise. 1989. *Aristotle on substance: The paradox of unity*. Princeton: Princeton University Press.
Goethe, Johann Wolfgang von. 1988. Scientific studies. In *Goethe, the collected works* (trans: Miller, D. ed.), Vol. 12. Princeton: Princeton University Press.
Gotthelf, Allan. ed. 1985. *Aristotle on nature and living things*. Pittsburgh and Bristol: Bristol Classical Press and Mathesis Publications.
Gotthelf, Allan. 1987. First principles in Aristotle's parts of animals. In *Philosophical issues in Aristotle's biology*, ed. Allan Gotthelf and James G. Lennox, 167–198. New York: Cambridge University Press.
Gotthelf, Allan. 1997. The elephant's nose: Further reflections on the axiomatic structure of biological explanation in Aristotle. In eds. W. Kullmann and S. Föllinger, pp. 85–96.
Gotthelf, A. 1999. Darwin on Aristotle. *Journal of the History of Biology* 32: 3–30.
Gotthelf, Allan, and James G. Lennox (eds.). 1987. *Philosophical issues in Aristotle's biology*. New York: Cambridge University Press.
Grene, Marjorie. 1972. Aristotle and modern biology. *Journal of the History of Ideas* 33: 395–424.
Grove, A.T., and Oliver Rackham. 2001. *The nature of Mediterranean Europe: An ecological history*. London: Yale University Press.
Harvey, W. 1993. *The circulation of the blood and other writings* (trans: Kenneth, J. F.). New York: Everyman Press.
Heidegger, Martin. 1995. *The fundamental concepts of metaphysics: World, finitude, solitude* (trans: McNeill, W. and N. Walker). Bloomington and Indianapolis: Indiana University Press
Holmes, R.L., J. Renn, and H.J. Rheinberger. 2003. *Reworking the bench: Research notebooks in the history of science*. Boston: Kluwer.
Hureau, J. C. 1986. Mullidae. In *Fishes of the North-eastern Atlantic and the Mediterranean*, World biodiversity database, CD-ROM, ed. J. C. Hureau.
James, P., and N. Thorpe. 1994. *Ancient inventions*. New York: Ballantine Books.
Jonas, Hans. 1968. The biological foundations of individuality. *International Philosophical Quarterly* 8: 231–251.
Judson, L. 1991. *Aristotle's physics: A collection of essays*. Oxford: Oxford University Press.
Kosman, L.A. 1987. Animals and other beings in Aristotle. In *Philosophical issues in Aristotle's biology*, ed. A. Gotthelf and J. Lennox, 360–391. New York: Cambridge University Press.
Kullman, Wolfgang. 1985. Different concepts of the final cause in Aristotle. In *Aristotle on nature and living things*, ed. A. Gotthelf, pp. 169–175.

Kullmann, W., and S. Föllinger (eds.). 1997. *Aristotelische Biologie*. Stuttgart: Franz Steiner.
Lange, Marc. 2004. The autonomy of functional biology: A reply to Rosenberg. *Biology and Philosophy* 19: 93–109.
Lee, H. D. P. 1985. The fishes of Lesbos again. In *Aristotle on nature and living things*, ed. A. Gotthelf, pp. 3–8.
Lennox, James G. 1983. Aristotle's Lantern. *The Journal of Hellenic Studies* 103: 147.
Lennox, James G. 1987. Kinds, forms of kinds, and the more and the less in Aristotle's biology. In *Philosophical issues in Aristotle's biology*, ed. Allan Gotthelf and James G. Lennox. New York: Cambridge University Press.
Lennox, James G. 2001a. *Aristotle: on the parts of animals*. Commentary (trans). New York: Cambridge University Press.
Lennox, James G. 2001b. *Aristotle's philosophy of biology*. New York: Cambridge University Press.
Lennox, James G. 2010. Unity and purpose of on the parts of animals I. In eds. J. G. Lennox and R. Bolton.
Lennox, James. 2011. Aristotle on norms of inquiry. *HOPOS: The Journal of the International Society for the History of Philosophy of Science* 1: 23–46.
Lennox, J.G., and R. Bolton. 2010. Being, nature, and life in Aristotle essays. In *Honor*, ed. A. Gotthelf. New York: Cambridge University Press.
Leunissen, Mariska. 2010. *Explanation and teleology in Aristotle's science of nature*. New York: Cambridge University Press.
Leunissen, Mariska, and Allan Gotthelf. 2010. What's teleology got to do with it: A reinterpretation of Aristotle's *Generation of Animals* V. *Phronesis* 55: 325–356.
Lones, T.E. 1912. *Aristotle's researches in natural history*. London: Newman & Company.
Luther, W., and K. Fielder. 1976. *A field guide to the Mediterranean sea shore*. London: Wm. Collins Sons.
Mayr, Ernst. 1963. *Animal species and evolution*. Cambridge: Harvard University Press.
Mayr, Ernst. 1982. *The growth of biological thought*. Cambridge: Harvard University Press.
Mayr, Ernst. 1996. The autonomy of biology: The position of biology among the sciences. *The Quarterly Review of Biology* 71(1): 96–107.
Miller, Jr, D. Fred, and Michael Bradie. 1984. Teleology and natural necessity in Aristotle. *History of Philosophy Quarterly* 1: 133.
Moraitou-Apostolopoulou, Maria, and Vassili Kiortsis. 1985. *Mediterranean marine ecosystems*. New York/London: Plenum Press.
Naddaf, Gerard. 2005. *The Greek concept of nature*. Albany: SUNY Press.
Nussbaum, Martha. 1978. *Aristotle's De motu animalium*. Text, commentary and interpretive essay (trans). Princeton: Princeton University Press.
Ogle, William. 1987. *Aristotle on the parts of animals*. Introduction and notes (trans). New York/London: Garland.
Pellegrin, Pierre. 1986. *Aristotle's classification of animals: Biology and the conceptual unity of the Aristotelian Corpus* (trans: Preus, A.). Berkeley: University of California Press.
Qugnard, J. P., and A. Pras. 1986. Labridae. In *Fishes of the North-eastern Atlantic and the Mediterranean*, World biodiversity database, CD-ROM, ed. J. C. Hureau.
Rackham, H. 1940. *Pliny. Natural history*, vol. III. Cambridge, MA: Harvard University Press.
Rheinberger, H.J. 1997. *Toward a history of epistemic things: Synthesizing proteins in the test tube*. Stanford: Stanford University Press.
Ritter, W.E. 1932. Why Aristotle invented the word entelechia. *The Quarterly Review of Biology* 7(4): 377–404.
Rosenberg, Alex. 2001. Reductionism in a historical science. *Philosophy of Science* 68: 135–163.
Schofield, M., M. Burnyeat, and J. Barnes (eds.). 1980. *Doubt and dogmatism*. New York: Oxford University Press.
Sedley, David. 2010. Teleology, Aristotelian and Platonic. In eds. J. G. Lennox and R. Bolton.
Shoja, Mohammadali, R. Shane Tubbs, Marios Loukas, and Mohammad R. Ardalan. 2008. The Aristotelian account of 'heart and veins'. *International Journal of Cardiology* 125: 304–310.

Sibum, H. Otto. 2000. Experimental history of science. In *Museums of modern science, Nobel symposium 112.*
Simpson, G.G. 1961. *Principles of animal taxonomy*. New York: Columbia University Press.
Straus, Erwin. 1966. *Phenomenological psychology*. New York: Basic Books.
Thompson, D'Arcy. 1913. *On Aristotle as a biologist*. Oxford: Clarendon Press.
Thompson, D'Arcy. 1936. *A glossary of Greek birds*. London: Oxford University Press.
Thompson, D'Arcy. 1942/1992. *On growth and form*. New York: Dover
Thompson, Darcy W. 1947. *A glossary of Greek fishes*. London: Oxford University Press.
Tipton, Jason. 1999. Darwin's beautiful notion: Sexual selection and the plurality of moral codes. *History and Philosophy of the Life Sciences* 21(2): 1–19.
Tipton, Jason. 2001. Division and combination of labor in Aristotle's biological writings. *The Journal of Bioeconomics* 3: 51–55.
Tipton, Jason. 2006. Aristotle's study of the animal world: The case of the kobios and phucis. *Perspectives in Biology and Medicine* 49(3): 369–383.
Tipton, Jason. 2008. Aristotle's observations on the foraging interactions of the red mullet. *Archives of Natural History* 35(1): 164–171.
Tipton, Jason. 2011. Borrowed plumes: Mimetic powers and the polymorphism of humans. *Biology & Philosophy* 26(6): 837–856.
Tweney, Ryan. 2004. Replication and the experimental ethnography of science. *Journal of Cognition and Culture* 4(3): 731–758.
Voultsiadou, Eleni, and Dimitris Vafidis. 2007. Marine invertebrate diversity in Aristotle's zoology. *Contributions to Zoology* 76(2): 103–120.
Wilson, Malcolm. 1997. Analogy in Aristotle's biology. *Ancient Philosophers* 17(2): 335–358.

Index Locorum

J.A. Tipton, *Philosophical Biology in Aristotle's Parts of Animals*, Studies in History and Philosophy of Science 26, DOI 10.1007/978-3-319-01421-0,

Parts of Animals (PA)

Subject Index

J.A. Tipton, *Philosophical Biology in Aristotle's Parts of Animals*, Studies in History and Philosophy of Science 26, DOI 10.1007/978-3-319-01421-0,

MIX
Papier aus verantwortungsvollen Quellen
Paper from responsible sources
FSC® C105338

If you have any concerns about our products,
you can contact us on
ProductSafety@springernature.com

In case Publisher is established outside the EU,
the EU authorized representative is:
Springer Nature Customer Service Center GmbH
Europaplatz 3, 69115 Heidelberg, Germany

Printed by Libri Plureos GmbH
in Hamburg, Germany